Lecture Notes in Electrical Engineering

Volume 92

For further volume
http://www.springer.com/series/7818

Antonio Liotta · George Exarchakos

Networks for Pervasive Services

Six ways to upgrade the internet

Springer

Antonio Liotta
Department of Electrical Engineering
Eindhoven University of Technology
P.O. Box 513
5600 MB Eindhoven
The Netherlands
e-mail: a.liotta@tue.nl

George Exarchakos
Department of Electrical Engineering
Eindhoven University of Technology
P.O. Box 513
5600 MB Eindhoven
The Netherlands
e-mail: g.exarchakos@tue.nl

ISSN 1876-1100
e-ISSN 1876-1119

ISBN 978-94-007-1472-4
e-ISBN 978-94-007-1473-1

DOI 10.1007/978-94-007-1473-1

Springer Dordrecht Heidelberg London New york

Cover design: eStudio Calamar, Berlin/Figueres

Printed on acid-free paper

Springer is part of Springer Science+Business Media (www.springer.com)

To Maria, Dikeos and Marina

Foreword

When I began working for Peter Kirstein's group at the University College London (UCL) in 1980 the department had already been part of the embryo Internet for seven years. The first "local area networks" were being deployed, and it was becoming clear that the future would consist of many networks, using a variety of technologies, all of which would need to interwork. Accordingly the Internet community adopted an architecture (designed by Robert Kahn and Vint Cerf) in which all networks must implement a common "internet protocol" (IP) to carry packets of data across and between networks. *Networks* were connected to their neighbours by computers termed "gateways" (we now call them "routers"). IP did not attempt to correct any errors that might arise; that was left to the "hosts"—the computers attached to the networks that were the sources and sinks of data. Hosts implemented a protocol called the "transmission control protocol" (TCP), which arranged for the re-transmission of any packets that did not arrive intact.

Back in 1980 routers were based on refrigerator-sized "mini-computers" which cost tens of thousands of pounds (hundreds of thousands of euros in today's terms). Connecting a computer to a LAN cost more than £1,000. However, by the time Antonio Liotta joined us at UCL in 1995 things were very different; costs had plummeted, personal computers were widespread and the Internet now comprised thousands of networks and millions of hosts. The applications that generated the bulk of the traffic though—file transfer, email and the burgeoning world wide web—still matched the requirements that had inspired the development of the TCP all those years before. Underlying TCP is the assumption of a client–server model; the server computer has something the client wants and the TCP delivers it complete and error free with high probability. TCP achieves this by trading timeliness for reliability. That is fine for applications like email—no one cares much if an email message is delayed for a few seconds provided it arrives intact. However, it was clear to Antonio and others researching in the late 1990s that new applications were on the horizon, many of which would not fit the TCP client–server model at all well. Some, such as streaming audio and video, would not tolerate TCP induced delays. Others were abandoning the asymmetric client–server model in favour of a more egalitarian "peer-to-peer" approach typified by file-sharing applications such as Napster. Yet another development, exemplified by Antonio's own research, turned the client–server model on its head by moving the servers ("agents") around the network in order to complete a task in the most efficient way.

Today those anticipated developments have arrived with a vengeance. Traditional TCP-based applications now form just a small minority of Internet traffic, perhaps no more than 15%, and most of that is world wide web. Streaming applications comprise around 10% and much of the rest is peer-to-peer file sharing—mostly of videos. (Things change so rapidly that in 12 months time these estimates will all likely be wrong!) Just as the applications have changed, so have the devices on which they run. The number of mobile phones in the world already vastly exceeds the number of networked computers and, increasingly, these phones themselves are on the Internet. Not only will there be hundreds of millions of them but they will move about! Mobility brings its own set of problems: wireless connections are subject to rapid changes in transmission speeds and error rates; an IP address is no longer a reliable clue as to where in the world a host is located.

Plainly the Internet has already adapted somewhat to support today's applications and host mobility. However, the adaptations have often been piece-meal, and stresses and strains sometimes appear. Researchers today must not only look at how better to adapt the Internet to today's applications but must also anticipate the huge changes that are, inevitably, around the corner. The authors of this book have, between them, accumulated many years researching novel techniques for optimising novel technologies within the Internet. They are well placed to understand the problems that must be solved and what solutions might be feasible. They begin by describing the key features of the Internet as it has evolved and the problems that must be addressed if it is to become flexible enough to support today's applications and mobility. They then look at what further adaptations may be needed within the Internet of the future. They do not make the mistake of claiming to know precisely what will be needed. Rather, they have used their knowledge to identify, as their subtitle makes clear, *Six ways to upgrade the Internet*. They explain these upgrades with the aid of carefully chosen examples and illustrations. The result is a book that will be of great benefit to students who wish to progress from an understanding of what the Internet is now towards an understanding of the motivations and techniques that will drive its future.

London, January 2011

Graham Knight
Department of Computer Science
University College London

Preface

Through the eyes of billions of Internet users, we have learned how the ease of communication can ignite phenomenal innovation. It is fascinating to witness the new habits and social phenomena created by the Web. However, what happens behind the scenes of our digital ecosystem? It is the network that moves our data around, handles the peak-hour traffic and strives to smoothly deliver the audio-video streams. Networks play a vital role in sustaining the unrelenting evolution of the most demanding Web systems.

Networks have to keep up with unprecedented data volumes while adapt-ing to new communication patterns or, rather, new kinds of traffic. Most applications are now *pervasive*. We expect them to be accessible everywhere, without compromise. We expect the same "look and feel," and the same quality and functionality, irrespective of any other technological constraints. Hence, many fear that the emerging breed of *pervasive applications* will soon render the Internet obsolete. As a matter of fact, a worldwide effort to reinvent the Internet is well underway by the "Future Internet" research community.

Through our active involvement in the investigation and teaching of network protocols, we have come to realize how difficult it is to grasp networking concepts that exceed the horizon of TCP/IP (i.e., the Internet protocol). When it comes to *advanced network protocols*, specialist literature abounds with creative proposals. Yet, very few protocols manage to step out of the laboratory and into the commercial world.

Perhaps our most ambitious task in writing this book was to extract a selection of remarkable ideas from the scientific literature and make them accessible to the non-specialist reader. Our book does not have the objective of embracing the *Future Internet*, though it does introduce a series of *network mechanisms* that will certainly find a place in the next-generation network. We propose six ways to upgrade the Internet and make it more *ubiquitous*, *reactive*, *proactive*, *information-driven*, *distribution-efficient* and *searchable*. In the final chapter, we offer some considerations about the Future Internet, though we have resisted the temptation to give any specific technical solutions.

This book is self-contained and is meant for anybody with an interest in the *post-Internet* era. We use the book to teach *ad hoc networks* and *P2P networks* in our *Communicating Systems* course at the Technical University Eindhoven (The Netherlands). You do not need to have a background in *computer networks* because all necessary concepts are summarized in the first two chapters. We have had to face the challenge of teaching *networking* to students who are not keen mathematicians: our efforts are reflected in this book which does not contain equations or mathematical formulations, but is enriched by examples and illustrations.

Yet, this is not another book for "dummies." Whoever has taken a classic course in *computer networks* will find our book to be a useful tool for gaining a deeper understanding of more *advanced network mechanisms*.

We hope that scholars in the field will find inspirational ideas within these pages for their research.

January 2011

Antonio Liotta
George Exarchakos

Acknowledgments

This book has been in our mind for several years, but it started as a concrete project only after we moved to the *Technical University of Eindhoven* (The Netherlands). At TU/e we found the inspirational energy that sustained our efforts. Several improvements came after discussions with our colleagues at *Electrical Engineering* and *Computer Science* and thanks to the interactions with our *Communicating Systems* students.

Springer was instrumental in bringing the book to the light of the day. We were lucky to work with an enthusiastic publishing editor who not only championed the book but also taught us a lot of things about writing for non-specialist readers. Many thanks to Rachel Hopkins who patiently copyedited the book in astounding detail.

We are particularly grateful to Lisandro Granville for injecting great ideas into the book. Graham Knight and Raouf Boutaba were the first to provide scientific feedback on the manuscript. Alessandro Liotta helped making Chaps. 1, 3 and 10 readable to the non-technologist.

On a more personal note, we are immensely grateful to our respective partners for supporting us throughout the writing process.

Contents

1 **On the Way to the Pervasive Web** 1
1.1 The Net, a Tool for Everyone 1
1.2 The Inexorable Transformation of Internet Applications 3
1.3 The Application's Mutiny 5
1.4 Everything on the Move 9
1.5 New Interaction Paradigms Emerge 10
1.6 The Scent of Pervasive Applications. 12
1.7 The Billion Dollar Question 13
References 14

2 **The Network, As We Know It** 15
2.1 The Multiple Facets of Networks 15
2.2 Networks from the Eyes of an Ordinary User 16
2.3 Invite a Programmer to Understand What's in the Cloud 18
2.4 A Network Engineer to Turn a Switch into a Router 20
2.5 The Computer Science of a Router. 23
2.6 Simple Math to Stabilize the Net 27
2.7 Life of a Commuter 33
2.8 The Three Fundamental Principles 35
References 38

3 **Six Problems for the Service Provider**. 39
3.1 The Net has Ossified 39
3.2 Problem 1: Not Truly Ubiquitous. 42
3.3 Problem 2: The Unresponsive Net 44
3.4 Problem 3: Too Much, Too Stale Signaling. 44
3.5 Problem 4: Lack of Parallelism 46
3.6 Problem 5: Data Agnosticism 48
3.7 Problem 6: Inadequate Net-Search Engine. 49
3.8 Concluding Remarks. 50
References 50

4 **Spontaneous Networks** 51
4.1 The Gift of Ubiquity 51
4.2 Spontaneous Connectivity 53
4.3 The Hidden-Terminal Problem 54
4.4 The Exposed-Terminal Problem 55
4.5 Preventive Measures to Avoid Collision 55
4.6 Path Discovery in a Volatile Networks 58
4.7 The KISS Approach 59
References 62

5 **Reactive Networks** 65
5.1 Why Networks on Demand? 65
5.2 A Traffic-Free Network 66
5.3 Our First Path 66
5.4 Path Management 69
5.5 Our Second Path 73
5.6 Global Synchronization 73
5.7 Error Management 75
5.8 Remarks on Reactive Networks 77
References 77

6 **Proactive Networks** 79
6.1 From Reactive to Responsive 79
6.2 Keep the Network Ready 80
6.3 How do I Find My Multipoint Relay? 81
6.4 Life of an OLSR Node 82
6.5 The Node's Information Repository 84
6.6 Shortest Path over the MPR Sub-topology 84
6.7 A Complete Example 85
6.8 How Proactive Can You Be? 87
6.9 The Power of Hybrid Protocols 90
References 93

7 **Content-Aware Networks** 95
7.1 Routers Should Read the Content 95
7.2 A Network on Top of the Physical Network 96
7.3 Centralized Assignment of Node Identifiers 99
7.4 Centralized Entry Point Discovery 99
7.5 Multiple Bootstrap Servers 102
7.6 Decentralized Assignment of Node Identifiers 104
7.7 Entry Point Discovery via Underlying Links 104
7.8 Content is an Asset at the Edges 107
References 108

8 Distribution–Efficient Networks 111
8.1 Publishing Goes Beyond Bootstrapping. 111
8.2 The Two Flavors of Virtual Networking 112
8.3 Creating Unstructured Neighborhoods. 113
8.4 Making Yourself Known in Unstructured Neighborhoods 116
8.5 Unstructured Resource Publishing 117
8.6 Secure a Role in Structure Worlds 121
8.7 Build Strict Formations. 122
8.8 Place Links and Resources into a Structured Ring 126
8.9 Data-Awareness via Protocol-Agnosticism. 129
References 130

9 Discovering Virtual Resources 133
9.1 Four Ways to Reach a Resource 133
9.2 Assessment of Discovery Mechanisms 134
9.3 Containing the Proliferation of Discovery Messages. 134
9.4 Blind Discovery for Unstructured Networks 135
9.5 Informed Discovery in Unstructured Networks 138
9.6 Discovery in Loosely-Structured Networks 139
9.7 Deterministic Discovery in Structured Networks 142
References 144

10 A Peek at the Future Internet. 145
10.1 The Fourth Networking Principle: Beyond Mere Connectivity 145
10.2 Internet of Things: Sense and Influence Your Environment. . . 146
10.3 Small, Large Networks 147
10.4 Manage the Autonomics 150
10.5 Dependable Networks. 150
10.6 The Fine Line Between Freedom, Security and Privacy 151
10.7 Energy-Efficient Networks 152
10.8 No Matter What, the Network will Remain Generative. 153
References 154

Index 157

Acronyms

AODV	Ad-hoc on-demand distance vector
APS	Adaptive probabilistic search
BFS	Breadth-first search
CIDR	Classless inter-domain routing
CPU	Control processing unit
CS	Client–server
CTS	Clear-to-send
DFS	Depth-first search
DHCP	Dynamic host configuration protocol
DHT	Distributed hash table
DiffServ	Differentiated services
DNS	Domain name system
DSL	Digital subscriber line
DV	Distance vector
FTTH	Fiber-to-the-home
HTL	Hops-to-live
IARP	Intrazone routing protocol
ICT	Information and communication technologies
ID	Identifier
IDS	Iterative deepening search
IERP	Interzone routing protocol
IntServ	Integrated services
IP	Internet protocol
IPTV	Internet protocol television
IS	Intelligent search
ISP	Internet service provider
KISS	Keep it simple and stupid
LAN	Local area network
LS	Link state routing
LSP	Link state packet
MACA	Multiple access with collision avoidance

MANET	Multiple ad hoc network
mBFS	Modified breadth-first search
MPR	Multipoint relays
NDP	Neighbor discovery protocol
NL	Neighbor list
NO	Network operator
OLSR	Optimized link state routing
P2P	Peer-to-peer
PC	Personal computer
PKI	Public key infrastructure
QoE	Quality of experience
QoS	Quality of service
RERR	Route error
RFIDs	Radio-frequency identification
RREP	Route reply
RREQ	Route request
RTS	Request-to-send
RW	Random walks
TC	Topology control
TCP	Transmission control protocol
TIB	Topology information base
TTL	Time-to-live
TV	Television
VoD	Video on-demand
VoIP	Voice over IP

Chapter 1
On the Way to the Pervasive Web

Abstract Web applications bring about extraordinary breakthroughs regarding our digital ecosystem. Pretty much anything with a chip and a radio interface can connect to the Web. However, many advocate a complete overhaul of the Internet as the only means to sustain innovation and productivity. Nobody knows what the next-generation of the Internet will look like; though important clues are visible as years of research have already generated phenomenal ideas. Together, we'll bring a range of network mechanisms "out of the lab" that can make the Net more proactive, reactive, robust and, ultimately, more pervasive than it is today. Our journey starts by scrutinizing the inexorable transformation of Web Applications in order to unveil the intrinsic limitations of the Internet.

The value of a network is proportional to the square of the number of communicating devices

Robert Metcalfe (inventor of the Ethernet)—1980

1.1 The Net, a Tool for Everyone

Just as in many other prominent technologies, networks exert a phenomenal impact on people. We don't need to look far to see how the Internet influences our daily routine. The Net's best incarnation, the World Wide Web, has become the number-one instrument we can't do without. Would you embark on a new trip without consulting "mother" Internet for weather and traffic information? Can you resist the temptation to seek a better fare online? Wouldn't you feel "naked" if you had to go through that important business meeting without the support of a

A. Liotta and G. Exarchakos, *Networks for Pervasive Services*,
Lecture Notes in Electrical Engineering, 92, DOI: 10.1007/978-94-007-1473-1_1,

Net-enabled laptop? Wiki fans, Twitters, and Facebook enthusiasts trust the Net with their personal life. Let us not forget the realm of entertainment, gaming, and video conferencing.

"Always-on" networks make marvelous things possible. Web dictionaries, e-encyclopedia, e-journals, e-books and all sorts of knowledge sources are readily and ubiquitously accessible. Thus, a "stable" data connection can make the life of a commuter less frenetic and even boost his productivity. As a matter of fact, this book would have never materialized if it weren't for dongles, hot-spots, Skype, WiFi-enabled trains and ... delayed trains—this is a commuter-generated book!

Indeed, the Net is not the first life-changing invention. Electricity, the radio, and the aircraft, for example, have made our lives brighter, more informed, and more mobile. However, the Net is by far a more stimulating instrument: in the hands of any ordinary user, it becomes "generative" [1]. The Net is a general-purpose "connectivity" machine. Its design doesn't constrain the user in any way. On the contrary, the variety of web applications that we see today has actually been facilitated by the Net's mechanisms.

Similar to other inventions, the Net hasn't changed much after its original conception. Although, the Net has a distinguishing feature: what we do with it (the applications) has changed enormously. Even more strikingly, the Internet architects never foresaw that video, audio and synthetic signals were all going to travel through a global data network.

At its infancy, the Net was not meant to be used by the general public. The growth rate of the last decade was unimaginable. A huge diversity has erupted in terms of applications, terminals and mobility patterns. The data that currently traverses local, metropolitan and wide-area networks is not of the same fabric as it was just 10 years ago. Networks are no longer just for email and browsing. They bear the load of file-sharing applications, Internet TV channels, video conferencing tools and gaming platforms.

The latest Web applications stress the Net in a new way [2, 3]. It's not merely the sheer volume of data: their anatomy is radically new. In former times when the only Net-killer was the email, data packets had a fairly relaxed life. Their delivery deadlines were in the order of seconds. Today's deadlines are in the area of milliseconds! We are dealing with the same network architecture, though with radically new usages trends. Yet, the Net is still functional, even several decades after its first deployment.

This book is about network mechanisms. An exploration of some remarkable research ideas reveals concepts that will make the Net more responsive, pervasive, and even more "generative" than it is today. Yet, before we can embark on this fascinating journey, we must meet two protagonists who are orchestrating the destiny of the Net: The Web developers, who add unprecedented innovation to the Net; and the Web users, with their escalating expectations that keep catalyzing inventions.

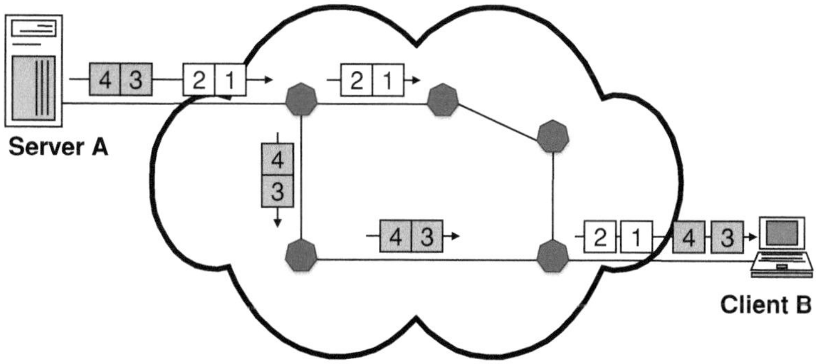

Fig. 1.1 A simple IP network

1.2 The Inexorable Transformation of Internet Applications

By its original conception, the Internet Protocol (IP) provides a simple yet universal mechanism for data packets (or datagrams) to find their way between terminals, even when those terminals are not directly attached to each other. However, IP networks can only offer a best-effort packet delivery service or, in other words, they don't have any effective means to [4]:

1. secure upper bounds on packet delivery times;
2. ensure that all packets (within the same session flow) follow the same path (from source to destination);
3. ensure that packets reach their destination in the same order they were sent;
4. guarantee that all packets are actually delivered.

Figure 1.1, in fact, shows a scenario in which a sequence of packets is sent from a server (A) to a user's terminal (B). IP networks divert traffic through alternative paths in order to avoid congested spots. This approach has proven to work extremely well for asynchronous or loosely-coupled applications where it's not crucial to keep source and destination in sync. Email and browsing are perfect examples of such a category. As we all know, there is much more to this in the Internet realm.

Figure 1.2 gives a qualitative view of the evolution of Internet applications, looking at four different dimensions. The first axis depicts the progression from "loosely-coupled" to "time-constrained" applications. The best effort nature of the Internet is a perfect match to email and browsing applications. As for the year 2010, Internet packets take (on average) 1–200 ms to reach their destination, and get lost with an approximate 8% rate (internettrafficreport.com). We can safely assume that the retransmission of lost packets incurs a further fraction of a second. All in all, an email might take, say, up to half of a second to be transported, i.e., an order of magnitude greater than the theoretical minimum. Congested email servers typically add precious seconds to the whole process. This means that the Net really is not always the bottleneck. Furthermore, who would complain about (or even notice) a 1-s email-delivery time? The same line of thought can be applied to online browsing.

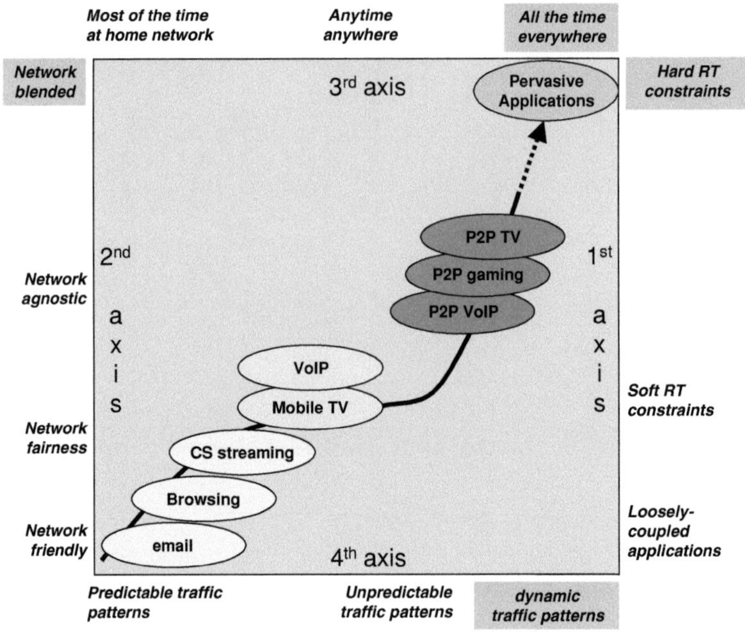

Fig. 1.2 The evolution of applications towards pervasiveness

Thus, our best-effort Net is perfectly geared to asynchronous applications that operate on timescales of the order of seconds. Streaming music and video stresses the Net with greater data volumes which have to be sustained for longer periods. The mere process of adding bandwidth to the core network can only help to a limited extent. The best-effort nature of the Net means that we still have packet loss and retransmission delays. On the other hand, a satisfactory quality of experience (by the watcher) demands a smooth video and audio delivery [5]. Application designers had to face a riddle: how to realize time-constrained applications on top of a network that doesn't acknowledge deadlines.

The Net behaves more like a lazy messenger than an Olympic athlete. Packets get delivered "just on time" and "most of the time." However, today we expect much more than email delivery—we expect record breaking packet delivery. The generative power of the Net has sparked a range of quick "fixes" (in jargon, "software patches") that make time-constrained applications tolerant to the Net's inefficiency. In fact, all those expedients have one important aspect in common: they operate on the terminals at the edge of the network. No modifications to the network core are allowed.

Perhaps the simplest and least ingenious trick is "edge buffering." Instead of playing-back packets as soon as they reach the receiver, we first reorder and buffer them on the user terminal. It is only when the size of the buffer gets longer than the typical network dynamics that we start the rendering process. Considering typical landline network conditions, buffering a few seconds worth of video will be sufficient to obtain a smooth video delivery.

Client–server (CS) streaming applications impose "soft" real-time constraints. Notoriously, edge buffering and various other media-encoding tricks help to absorb Net-related anomalies only up to a certain point. Lost packets will have to be recovered within specific deadlines to prevent Quality of Experience (QoE) degradation. Trade-offs are possible in order to tackle congestion, e.g., to increase buffer size. However, not everybody is prepared to accept lengthy buffering times.

Following the sparkling success of video streaming services, the next challenge was to deploy telephony over the Net—in jargon, Voice over IP (VoIP). In VoIP, we can still use edge buffering, but delivery-time constraints become much harder. Normally (and unless one is being patronized by a tenacious speaker), a phone conversation is a bidirectional communication process that becomes ineffective if the end-to-end delay becomes greater than, say, half of a second. Thus, we have a rigid upper bound on buffering time. Even worse, audio is less tolerant to packet loss and jitter than video.

Once it was established that the Net could sustain both TV and telephony, the step towards IP video telephony was short. Then, the challenge became keeping audio and video in perfect synch, which resulted in further tightening of packet delivery deadlines.

As new applications come to light, it's clear that the trend is for harder and harder real-time constraints. For instance, online games involve virtually unlimited folks acting on the same scene, which has to be rendered consistently across the board. However, different application components might have distinct requirements. Social networking platforms come with the ability to embed a myriad of applications, ranging from loosely to tightly-coupled components. A "Tweet" (in Twitter.com) and a "Wall Message" (in Facebook.com) are less time-constrained than a "textual chat." This is less critical than playing a video clip. However, as we enter the realms of multi-party video conferencing or augmented-reality games, we begin to push the Net beyond its limits.

1.3 The Application's Mutiny

In the IP world, we can say that loosely-coupled applications are the good guys, as they conform to the "best-effort" service concept. From another angle, we can also say that the Net has actually been designed to multiplex the traffic of best-effort applications. Thus, email and browsing are in the category of "network-friendly" applications (Fig. 1.2, 2nd axis).

When client–server streaming applications became popular, the issue of "network-fairness" arose. As more people trigger time-constrained packets, there is a risk that a few voluminous flows will take over the Net. Protocols such as TCP (Transmission Control Protocol) offer some (rather limited) means to encourage a fair sharing of network capacity. However, the Net is not designed to police fairness and, today, "bandwidth-hungry" applications have become commonplace.

If we observe what has been happening in the last few years alone, there is a clear divergence between application requirements and what the network can actually sustain. In fact, there is a clear trend towards "network-agnostic" applications. These operate under the assumption that the Net is always able to (somehow) provide a "bit-pipe" from any point to any other point. Hence, the application developer has set himself free from the need to make compromises with the Net, focusing on the fabrication of new interaction paradigms.

The step from "network-friendly" to "network-agnostic" applications represents the apogee of the Net's generative power. It has led to outstanding developments, even original forms of communication paradigms—new ways in which people interact. The phenomenal impact of social networking, user-generated content sharing and virtual-collaboration tools is well known. Today, the multitude of free online tools exerts an unprecedented social influence. People announce their wedding on Facebook and, then, share their best honeymoon moments online. Thanks to recommendation platforms such as TripAdvisor, sloppy hotel managers are publically "named and shamed" by frustrated customers. Ordinary people sell virtual objects in Second Life, giving new breath to their business creativity.

The "social network" revolution has also brought inconceivable distress to some incautious people. A teacher can lose her job for simply publishing a single "unorthodox" photograph online. Successful companies can be put out of business by unscrupulous competitors who can play dirty with the eBay reputation system.

Let us set aside the marvels and nightmares of the Web phenomenon in order to return to the Net. Let's pick a prominent example of a network-agnostic application and examine its impact on the Net. Peer-to-peer TV (P2P TV) adopts a clever trick to distribute TV content over the Net on a massive scale [6]. Instead of using individual channel servers which have limited capacity, P2P TV platforms use the viewer's computers to relay the stream to other viewers.

Figure 1.3 gives a first impression of a typical P2P TV system [7–10]. In this particular scenario, we assume that the stream originates from the server of a broadcasting company (Fig. 1.3a). Other systems actually allow user-generated content, though are not substantially different. Once the stream has reached a number of user terminals, these keep a copy of the stream in store and form a distribution network (Fig. 1.3b, c). Those terminals can both "stream in" and "stream out," and are thereby termed *peers*.

What are the benefits of this approach? Thanks to the collaboration among peers, P2P TV avoids the concentration of load around the TV broadcasting server. The P2P distribution mechanism is considerably more scalable than its client–server counterpart. From the user's perspective, P2P TV is certainly a good invention.

Let's change our approach: what do Internet Service Providers (ISPs) and Network Operators (NOs) have to say about P2P traffic? Does P2P TV make an economic use of the Net? Scientific rigor would involve an extensive scrutiny—new P2P TV platforms come out like "mushrooms in September." Nevertheless,

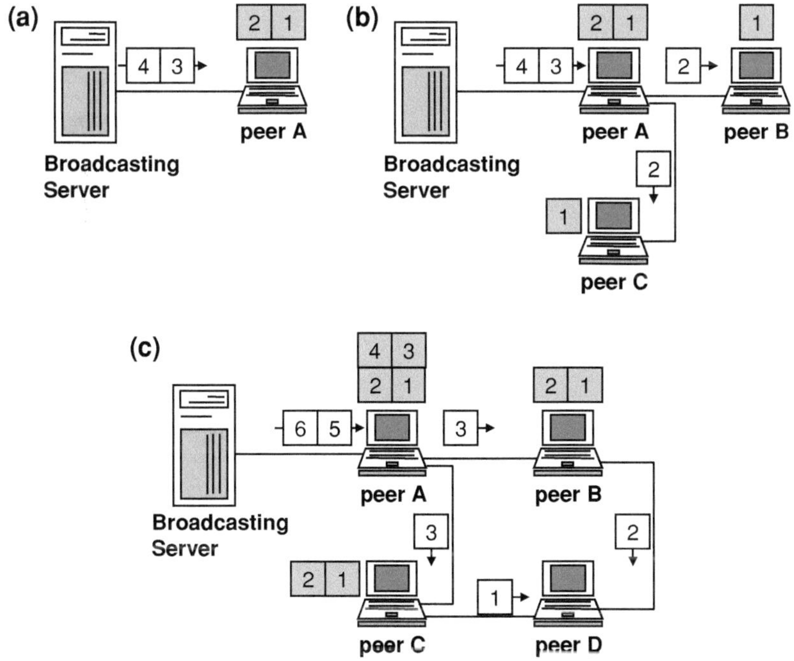

Fig. 1.3 Peer-to-peer television over a wireline IP network. **a** Peer A has buffered chunks 1 and 2 of the video stream. **b** Peers B and C obtain the first two chunks from peer A, rather than resorting to the broadcasting server. **c** Peers B and C get subsequent chunks from peer A. Peer D gets chunks 1 and 2 in parallel from peer C and peer B, respectively

for the sake of simple argumentation, let's catch a glimpse of Joost (www.joost.com).

Figure 1.4a gives a snapshot of the distribution of traffic sources and destinations recorded for 1 h in 2009 [11]. The test was carried out in a laboratory located at Colchester University, United Kingdom. Thus, one would imagine that most of the traffic would hang around the southeast of the UK. By contrast, the test revealed that only 21% of traffic originated within the UK boundary, with a good portion coming all the way from Canada (15%) and South Africa (10%). This is not good news for ISPs, as the cost to transport bits across continents is considerably high.

The bad news doesn't end here. The P2P paradigm involves more than just "getting" content. It's also about "contributing" content. Our test-site gathers content from some computers and, at the same time, relays stream chunks to other TV viewers. Imagine a child watching "Felix the Cat" in Essex, UK. According to Fig. 1.4a, b, there is a good chance that the stream comes in from Canada (15% probability) and is relayed out to the computer of another child who is sitting in Australia (23% probability).

Intuition suggests that it would be better to keep the stream within the UK border. Yet, like many other P2P platforms, Joost strives for "scalability" more

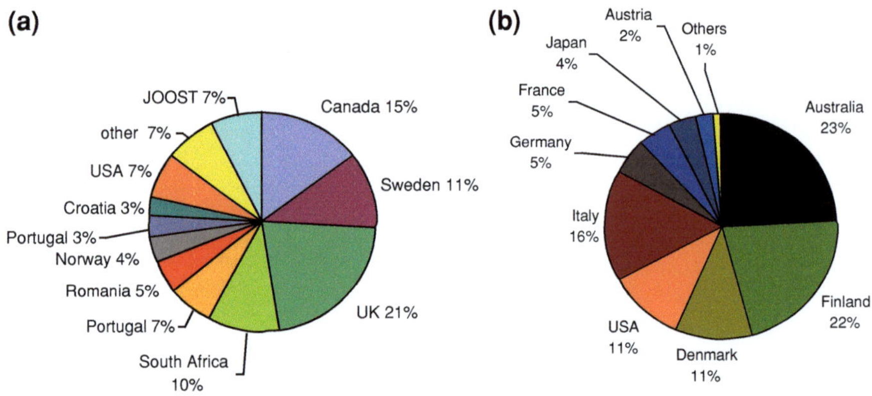

Fig. 1.4 Network agnosticism in a popular P2P TV platform. Geographic distribution of **a** traffic sources and **b** destinations (*Source* [11])

than it pursues network efficiency. If the platform put restrictions on the choice of inter-communicating peers, we would end up with localized clusters of peers, whereas the maximum scalability in P2P is achieved on a global scale.

Achieving a sustained quality of the TV stream is not straightforward when the network is "best-effort" and the peers are "volatile." If the child is streaming "Felix the Cat" from a Peer sitting in Canada, what happens if, halfway through the program, the source computer is switched off? An effective strategy is to get the stream from multiple peers. Therefore, if one source disconnects, we don't lose the channel.

We can't predict when peers disconnect, though we can play a statistical game. We keep the set of peers as large as possible; we always stream from multiple peers; and we keep changing the sources. For this approach to work smoothly, we must distribute the traffic as much as possible [12].

This simple P2P TV example underlines the increasing divide between those who design Web application and the companies that operate the Net. Here is some food for thought thus far:

1. Emergent applications are evolving into the realm of new interaction paradigms that are not always "network friendly."
2. The other perspective is that networks don't offer "native" support to new communication paradigms.
3. History tells us that after the IP network revolution, networks have followed an "evolutionary" path, whereas applications continue to experience "revolutionary" steps.

Thanks to the "generative" feature of the Net, Web developers may soon come up with even more revolutionary applications. Yet, the best-effort Net might not be able to sustain them. Our best guess for the future is that the next-generation of Web applications will "blend" with the Net. Perhaps the next paradigm shift will surpass the hurdles of the current "network-agnostic" Web.

1.4 Everything on the Move

Thus far, we have examined the emergent Web applications and how these tend to diverge from the existing capabilities of the Net. Web users are the other important protagonists. Our expectations from the Net keep rising [13]. We want to embrace the power of the Web when we are in the office, at home and while we move from one place to the other. However, the Net was not geared for mobility.

The original IP world was absolutely stationary. In the 1990s, one would have to obtain a unique IP address and machine name before joining the Net. The IP address was bound to a specific network access point. Thus, if you wanted to move a computer from one building to the next, you had to obtain a new IP address. Stationary IP networks are topologically more static. Hence, it's much easier to stabilize them.

Wireless and cellular networks make the Net substantially more dynamic and transient. Mobile networks extend the early IP without changing its design. They "patch" the Net with edge mobility without improving the core architecture. This "evolutionary" (patching) approach seems to work satisfactorily for loosely coupled applications. Relaxed delivery deadlines allow sufficient time for the Net to keep track of mobile terminals.

On the other hand, time-constrained applications and the P2P interaction paradigm strain the mobile Net to the utmost. Figure 1.5 depicts a "mobile" CS streaming scenario to exemplify the issue. In snapshot Fig. 1.5a, a streaming session is well underway. After the terminal handovers to a new access point, the network takes some time to re-compute the new destination. A plain IP Net would take anywhere from minutes to hours to detect that Client B has a new IP address. This would leave Client B with no other choice but restarting the streaming session from scratch. More hints on "patches" that can accelerate handover will be given in Chap. 2. However, due to intrinsic design limitation, a plain IP network won't be able to perform instantaneous handovers. Thus, packets will always be lost in the process, as sketched in Fig. 1.5b.

With the P2P interaction paradigm, mobility becomes even more problematic. By contrast to the CS scenario (where at least the servers stay put), with P2P, everything can potentially move (both the "data source" peer and the "data sink" peer). Keeping track of mobile peers is a nightmare; it's practically impossible with current technologies.

As of 2010, network technology is just not able to combine mobility with the P2P paradigm (genuine mobile P2P TV is a futuristic concept). The same is valid regarding time-constrained web applications. Thus, today we have already reached a severe limiting factor that might delay the invention of new Web paradigms.

On the other hand, the digital ecosystem of the "always-connected" society demands for ever-increasing degrees of mobility (third axis of Fig. 1.2). As of today, we have gone past the era when people were constrained to landline access. High-speed cellular networks provide a fairly stable connection in the most inhabited areas—provided that we do not travel too fast. Cellular networks are

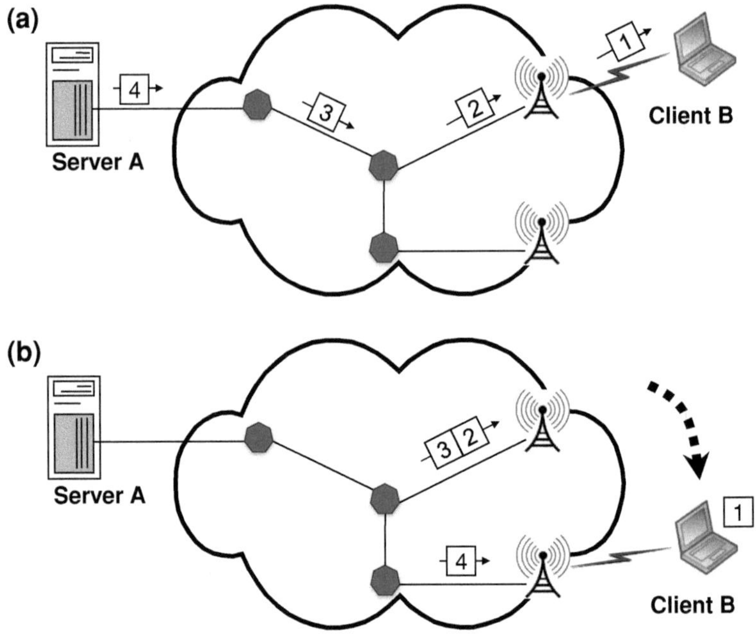

Fig. 1.5 Handover between different networks. **a** Client B has established a streaming session with the server. **b** The client's handover causes a session break up. Packets 2 and 3 which were on their way to the previous access point are deemed to be lost

getting better, although ubiquitous access is still a chimera. Yet, even if 1 day we reach the goal of "always-on-everywhere" connectivity, the structural limitations of the IP mentioned above will still have to be overcome.

Eminent technologists, researchers and policy makers are now facing a daunting dilemma: is the Internet going to sustain the combined strain of mobility and real-time applications? In other words: will the "Internet patching" Internet patching approach suffice or do we urgently need to re-think the Internet?

1.5 New Interaction Paradigms Emerge

The 4th axis of Fig. 1.2 identifies another issue that has to do with the dynamics of traffic patterns in the Net [14]. In the original IP design, the Net merely ships one packet after the other, from source to destination. The main concern is to keep the network core as simple as possible. Thus, the Net is dumb machinery that treats packets as one would treat hot potatoes: the priority is to get them out of your hands.

In the scenario of Fig. 1.6a, the Net is unaware that two users, Clients B and C, are getting the same sequence of packets (in jargon, "data flows"). The Net has no

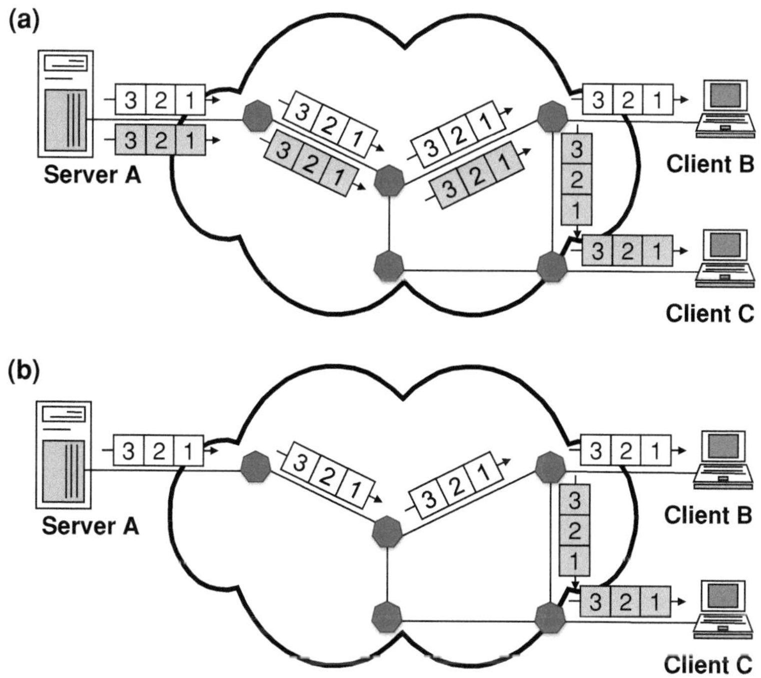

Fig. 1.6 Stream distribution mechanisms in an ordinary wireline IP network. **a** Multiple unicast transmission; **b** Multicasting transmission

means to detect the unnecessary duplication of packets. In fact, a much more efficient distribution mechanism would be that of Fig. 1.6b, where packets are duplicated only when strictly required. Yet, to the eyes of the Net, a packet is just a collection of bits. It's not possible to detect transmission patterns that would help to optimize the entire process.

Further along in the book, we'll see several networking tricks to improve the efficiency of data distribution. In essence, if we can determine the relative location of clients and servers, we can also build optimal distribution paths such as the one of Fig. 1.6b.

For now, let's get back to the issues of the emergent applications. The ability to engineer network mechanisms that can take advantage of usage patterns such as that of Fig. 1.6 is considerably diminished by user mobility. The data distribution path will have to be recomputed every time Client B or C moves. Obviously, the better the network can track (or even predict) the location of clients, the easier it will be to optimize the stream-distribution tree.

As applications become more network-agnostic applications, their traffic patterns become even subtler. Recall what was found about Joost and its statistical handover: traffic sources are continuously forced to change in order to spread the load among peers. This leaves the Net with a genuine disability: a network that can't recognize its own usage patterns will have a very limited ability to cope with

P2P applications. The trend is for very dynamic traffic and usage patterns: bad news for the IP architects.

1.6 The Scent of Pervasive Applications

The word "pervasive" is central to this book; so what's a pervasive application? Let's borrow a couple of definitions from the dictionary to seek a first sense of inspiration. "Pervasive: spreading widely through something" (Oxford Dictionary) and "having the quality or tendency to pervade or permeate" (Dictionary.com). Web applications reflect these definitions—they spread through the Net. Thus, by the rigor of logic, we must conclude that an application can truly manifest its "pervasiveness" only if the Net itself (the vehicle that brings the application to its user) is "pervasive."

This is still vague and abstract. We need a more pragmatic proposition. Within the more specialist literature, we find "pervasive computing," "pervasive networks" and "pervasive services;" but the scientific community has yet to agree about a common definition. Is this perhaps a sign that we are at the fringe of a technological revolution?

When scientists have a feel for something that is looming, they work around themes, properties and keywords. The most frequent ones found in specialist articles include: "seamless," "ubiquitous," "adaptive" and "context-aware." Currently, Web applications are "network-neutral", they work on top of any network, be it a fixed, wireless, cellular or satellite Net. As for the "network agnosticism" of P2P TV and other similar applications, these work "on top of" the Net, not "within" the Net. They exert a phenomenal impact on the Net, which is far from being "seamless." Pervasive applications should work in tandem with the Net instead of contrasting its operation. Thus, looking at the *status quo* of Web applications, there is significant scope for improvement. One day pervasive applications will "seamlessly" blend with the network. The two will be in such a harmony that they might even become indistinguishable.

The second property we picked was "ubiquitous." According to the dictionary, this is a synonym of "everywhere." Web applications meet this definition if we are content with the level of reach ability of the Internet. Today, provided that we have a decent computer and Internet connection, we can reach a server "most" of the time. Streaming quality is "mostly" acceptable if we agree to downgrade our expectations now and then. If the quality is not good enough, we just try again later. The whole infrastructure offers best-effort services so we can't get a truly ubiquitous experience.

Pervasive applications must be available all the time, everywhere, without compromise. Users must get the same "look and feel" regardless of which terminal or network access point they are using. The network will provide "adaptive" real-time constraints in the sense that it will understand the overall user's context (context awareness) and assist the application in delivering a uniform quality of experience (QoE).

If ubiquitous accessibility is an important property, then we also have to live with the idea that the traffic patterns of pervasive applications will be erratic. Today's networks are simply not able to make sense of what happens at the application level and don't have the means to offer a ubiquitous quality of service (QoS) when traffic sources and sinks are unpredictable.

In the journey towards pervasiveness, the destinies of applications and networks are intertwined. The four dimensions of Fig. 1.2 are spiraling towards unpredictable heights. Perhaps, completely new kinds of applications might appear if the next-generation network is to be intrinsically pervasive. Therefore, when we think about future networks, we shouldn't merely look at present applications [1]. In fact, policy makers and scientists across the world are now working on the assumption that current networks are limiting the ability to generate new applications and, with it, new business opportunities.

1.7 The Billion Dollar Question

If we had the ability to re-invent the Internet and replace it overnight, what would the next-generation network look like? The European Seventh Framework Research Program alone has allocated over €0.5 billion to find an answer to this question [15]. Forty-six research projects (for a total of €200 M) have started during the first quarter of 2008. Similar initiatives are well underway across the world (i.e., FIND and GINI in the US).

The next decade promises to deliver multiple, creative and (why not) conflicting answers to this fundamental question. Though, we'll have to wait a bit longer for the new recipe, the new Internet and the new catalyzers. Until then (and it might be a long wait), we thought it would be more constructive to reflect on some remarkable research ideas, bringing them out of the lab. In this book, we make a selection of noteworthy network mechanisms which have the potential to play a role in future networks. The aim is to familiarize the reader with methods that go beyond the plain IP world.

At the time of writing, there was a frantic attempt to forecast what the Future internet will look like, what it will and will not do. Company evangelists talk and write about our prospects, how the future Net will manage to support the needs of the ubiquitous user. Nobody holds "the" answer yet. Though fortunately, years of research in communication and networks have generated phenomenal ideas. Vital clues are already around us. Together, we'll unveil some of those clues, revisiting concepts and network mechanisms that will probably make it to the future. We'll see how to build networks that don't need any infrastructure; how networks can become more proactive, more reactive and more robust.

Our journey towards the "Pervasive Web" starts from the Internet. Chapter 2 catches a glimpse of the fundamental mechanisms that govern the Net, bringing relentless progress to our digital ecosystem. We'll revisit the mechanisms of the Net with a critical eye in order to better understand its intrinsic limits. Only then, will we be ready to explore mechanisms that can further enhance the pervasive nature of the Net.

References

1. Zittrain J (2009) The future of the internet and how to stop it. Yale University Press, New Haven
2. Liotta A, Lin L (2007) The operator's response to P2P service demand. Commun Mag IEEE 45:76–83
3. Pavlou G, Flegkas P, Gouveris S, Liotta A (2004) On management technologies and the potential of web services. Commun Mag IEEE 42:58–66
4. Tanenbaum AS (2002) Computer networks. Prentice Hall, Upper Saddle River
5. Agboma F, Liotta A (2009) QoE in pervasive telecommunication systems. Pervasive computing: innovations in intelligent multimedia and applications. Springer, Berlin, pp 365–382
6. Alhaisoni M, Liotta A, Ghanbari M (2009) Improving P2P streaming methods for IPTV. Int J Adv Intell Syst 2:354–365
7. Antonopoulos N, Exarchakos G, Li M, Liotta A (2010) Handbook of research on P2P and grid systems for service-oriented computing: models, methodologies, and applications. Information Science Publishing, Hershey
8. Liotta A, Antonopoulos N (2008) Computational P2P networks: theory & practice. IEEE Computer Society, Los Alamitos
9. Liotta A, Antonopoulos N, Exarchakos G, Hara T (2009) Advances in peer-to-peer systems 2009. IEEE Computer Society, Los Alamitos
10. Liotta A, Antonopoulos N, Kambayashi Y (2010) Advances in peer-to-peer systems 2010. XPS Publishing
11. Alhaisoni M, Liotta A (2009) Characterization of signaling and traffic in Joost. Peer-to-peer Netw Appl 2:75–83
12. Alhaisoni M, Ghanbari M, Liotta A (2010) Localized multistreams for P2P streaming. Int J Digit Multimedia Broadcast 2010:1–13
13. Agboma F, Liotta A (2007) Addressing user expectations in mobile content delivery. Mobile Inf Syst 3:153–164
14. Böszörmenyi L, Burdescu D, Davis P, Stanchev P, Fotouhi F, Liotta A, Newell D, Schöffmann K (2010) Advances in multimedia. IEEE
15. Papadimitriou D (ed) (2009) Future Internet—the cross-ETP vision document, v. 1.0. http://www.futureinternet.eu

Chapter 2
The Network, As We Know It

Abstract If you aren't a network guru and have no interest whatsoever in becoming one, though still wonder how the 'Net' works, this chapter will provide you with a number of precious answers. Together, we'll revisit the hectic journey of a data packet from the time it's conceived by an Internet server, until its destiny is accomplished in your computer. As with messengers, packets carry valuable information. Their purpose in life is simple: to find the best path to their addressee. However, in a network entangled with billions of links, how does your packet find its way through? How can streams of packets be delivered on time? You will appreciate the mechanisms that keep the network connected and stable. To those who are not network specialists, this chapter will provide all the elements required to tackle the more advanced networking concepts introduced in the rest of book. You will read about routers, packet switching, data buffering, message forwarding, the wonders of Dijkstra's algorithm and the tricks used to keep mobile terminals connected.

What happens depends on our way of observing it or on the fact that we observe it

W. Heisenberg, theoretical physicist (1901–1976)

2.1 The Multiple Facets of Networks

Networks and applications are complex systems where numerous actions and responses take place. Digital objects, components and alarms communicate, interact and interfere in the virtual world. Remote events are often interlaced: for instance, a web application, a mobile browser, and an IPTV client all rely on servers and become unresponsive when those servers are unreachable or overloaded.

A. Liotta and G. Exarchakos, *Networks for Pervasive Services*,
Lecture Notes in Electrical Engineering, 92, DOI: 10.1007/978-94-007-1473-1_2,

However, we have all experienced how the communication patterns and, with them, the complexity of the latest web applications are changing. A mixture of synchronous, asynchronous and virtual interaction paradigms characterizes the emerging social networks. Different people use Facebook (facebook.com) or Second Life (secondlife.com) in entirely different ways. The same applies to P2P file- or stream-sharing systems such as BitTorrent (bittorrent.com) or Joost (joost.com). New web applications relentlessly emerge every day. New collaboration and interaction paradigms will materialize. The strain on the network is inevitably becoming unbearable.

With these premises, it's no surprise that the Internet, the Web and their applications have grown into one of the most complex engineering beasts. Before we can think about the future of the Internet, what we would envision it to look like (Chap. 3) and how we might get there (from Chap. 4 onwards), we should revisit the current scenario.

Internet systems are complex; they involve network operators, service providers, broadcasting enterprises, application developers. They require varied and specialist skills to develop system components including hardware, software, middleware and underware (the latter not to be confused with French lingerie). Bookshops are inundated with relevant literature for professionals, consultants, programmers and ordinary people. Is it, nevertheless, possible to understand the essential features, mechanisms and deficiencies of the everyday Internetworked systems? Chapter 2 takes on this challenge.

We can't follow the classic engineering textbook approach [1–3], which would certainly require an encyclopedic *tour de force*. Let's try a new method based on perspectives and granularities. Networks are multifaceted. Depending on who is looking at them, a diversity of features arises. Ordinary users don't (and wouldn't want to) see the internals of network protocols. Mostly, they care about gaining access to their web applications and experiencing a high-quality service. A smooth quality of experience (QoE) is the result of multiple factors belonging to separate technical domains, ranging from physics (the communication channel) to control (channel sharing policies), but also management (network path and topology), programming (application design), electronics (capability of the user's terminal) and usage patterns (what people actually do on the Net) [4]. In this chapter, we revisit some of these facets and thereby identify three fundamental principles that govern current networks. Towards the end, you will wonder whether those principles are now starting to crack under the increasing pressure generated by the latest Internet applications.

2.2 Networks from the Eyes of an Ordinary User

The beauty of the Internet is that any ordinary user is able to use it productively. It isn't necessary for people to understand its internals. There is no need to know that the information travels across in the form of data packets (collection of bits

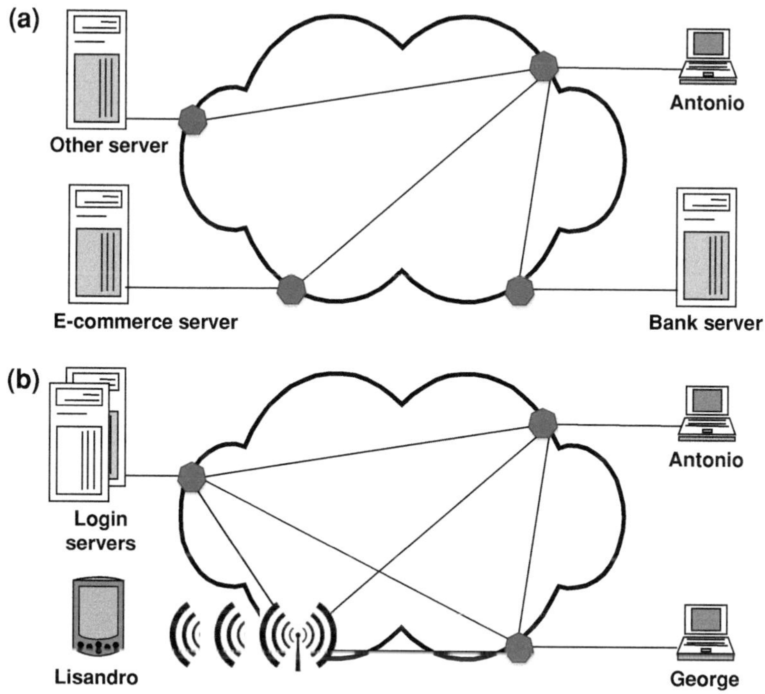

Fig. 2.1 A network seen from the eyes of an ordinary user. **a** In the Client–Server mode of operation, people have the impression that they are directly connected to any of the public Internet servers. **b** Using P2P programs, people have the impression of a full-mesh connection with anything on the Net

and bytes); one must not realize how those packets find their way from a server to your computer and the other way around. The browser and a few search engines connect you with the right server, application and data.

Depending on the type of web application they wish to employ, Internet users will see the Net as schematized in Fig. 2.1a or b. respectively. In the case of Client–Server (CS) applications [5], the interaction is mostly between human and machine, the web browser and one or more servers (a). To buy books online, Antonio fills up the basket of an e-commerce site. This will, in turn, redirect him to the secure transactional payment server of his bank. Being an ordinary user, Antonio won't need to know precisely how his credit card details are encrypted and routed to the bank's server and, then, relayed to the e-commerce site. If Antonio is in a phase of his life where he still trusts banks, all he cares about is a confirmation email which includes the transaction details.

The time of mere human-to-machine communication is over. Millions of people use the Net in more inventive ways. People communicate directly with little or no mediation via servers. Lisandro is again on the move between Brazil and Ecuador; Antonio is visiting his colleagues in Enschede, on the border between the Netherlands and Germany; George is back in Greece for a family reunion. How are they going to continue writing their book? Not a problem! There is broadband and a few P2P programs. They can directly communicate, for instance, via the video conferencing program Skype (www.skype.com). Following an authentication procedure (login servers), audio and video communication is server-less (Fig. 2.1b). Windows Live Synch (sync.live.com) supports shared document editing; several other programs will provide an online whiteboard plus incremental backups.

Antonio, Lisandro and George can make a productive usage of this fully meshed network. From their viewpoint, everything is connected to everything else. However, can you imagine a network connecting billions of people via dedicated, direct links? It is certainly not a viable proposition.

2.3 Invite a Programmer to Understand What's in the Cloud

Just like telephone networks, computer networks rely on switching devices to enable communication between terminals that are not directly connected (Fig. 2.2a). The billions of computers attached to the Internet are just a few hops away from each other thanks to a core switching and forwarding network. George connects to a remote IPTV service in order to watch the Sunday football highlights: a video stream is packetized on the server. Then, each packet has to find its way to George's terminal (Fig. 2.2b) where the stream will be reconstructed and rendered on screen.

However, George is not alone in this world; chances are that many more people will require those switches to forward other packets. Switches have finite capacity; links have finite bandwidth. Thus, there is a concrete possibility that at some point, as more users share portions of the network, resource contention will become an issue. While George is still busy watching football, Lisandro initiates a Skype video session with Antonio (Fig. 2.3a). George streams at high definition (packets 1–3), filling up the capacity of the S_b–S_c link. Hence, Lisandro's packets (4–6) have no other option than queuing up on the buffer of S_b (Fig. 2.3b), with detrimental consequences for his chat with Antonio.

Packet switching, buffering and forwarding from source (e.g., server) to destination (e.g., user terminal) are the fundamental mechanisms that keep people connected on the Internet. The applications that run on our computers can't ignore the fact that, because network links and switches are "shared" resources, sooner or later, contention will affect the packet delivery process.

Contention at switch S_b will cause packets 4–6 to be held in the switch's buffer. Buffering tries to keep all active connections alive; though, it causes other troubles. End-to-end communication latency increases. When packet delivery takes more than a few dozen milliseconds, Lisandro's video frames will be cluttered with pixelation. If the contention doesn't quickly subside, S_b will soon run out of buffer space and be forced to discard packets. Now, entire frames will have to be skipped; the video loses its smoothness and eventually goes out-of-sync with the audio.

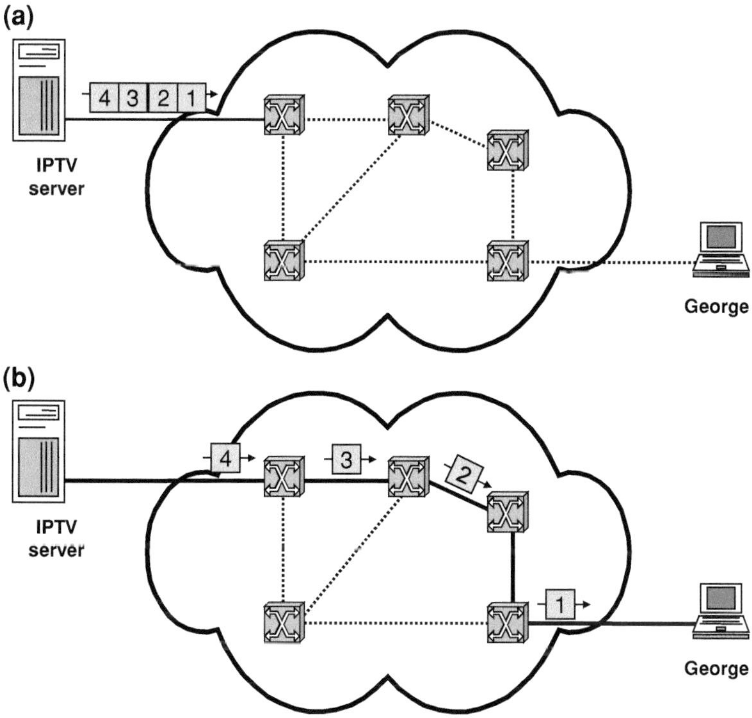

Fig. 2.2 Switching of data packets

Further delay and packet loss will finally affect the audio quality and lead to a communication breakdown.

Application developers must design Internet applications that are resilient to communication latency (end-to-end delay), latency variations (jitter), packet loss, and out-of-order packet delivery [6]. Special protocols such as the Transmission Control Protocol (TCP) operate on the network edge (i.e., run on the terminal) to keep senders and recipients in synch, re-order packets and demand the retransmission of dropped packets [1]. However, all these operations increase transmission latency from one end to the other. Also, when packets are dropped due to excessive congestion, retransmitting them is not always the best thing to do.

Depending on the type of application, tolerance to communication impairments may only be achieved to a limited extent. IPTV and video-on-demand (VoD) services are handled more easily [7]. First, a good chunk of the video stream is buffered onto the computer. Then, playback is realized locally. These applications have found a way around network contention problems, though they are prominently bandwidth-hungry.

Applications such as video conferencing are more problematic [8]. In this case, it's imperative to meet precise packet delivery deadlines. This is getting increasingly difficult because the Internet is a best-effort network. There are no

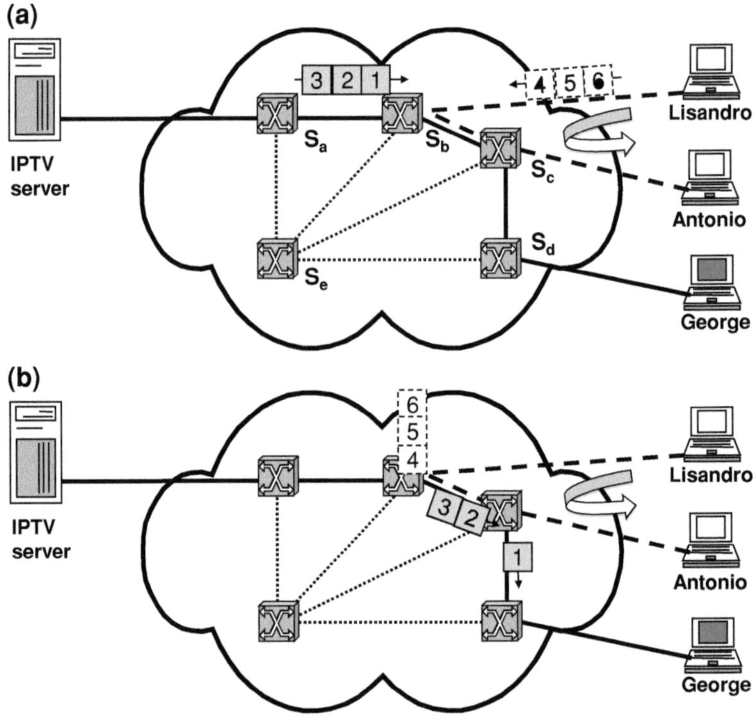

Fig. 2.3 The resource contention problem

mechanisms to enforce specific packet delivery deadlines. There aren't even guarantees that all packets are actually delivered. All in all, this approach has been working fine until now because the Net has been designed with redundancy in mind—any two points are reached via multiple paths. However, are things getting worse? At the time when this book went to press, the typical end-to-end transmission time was in the range of 120–140 ms; the typical packet loss rate was in the order of 8–10% (internettrafficreport.com). As new services push multimedia streams through this best-effort infrastructure, the situation is bound to deteriorate. Application programmers can no longer avoid working in tandem with network designers and multimedia specialists to overcome the challenges posed by the emerging bandwidth-hungry services.

2.4 A Network Engineer to Turn a Switch into a Router

What is inside a switch and how can this remarkable device help to distribute packets more evenly across the network? As with ordinary light switches, network switches need a trigger to change their status. Each and every Internet data packet

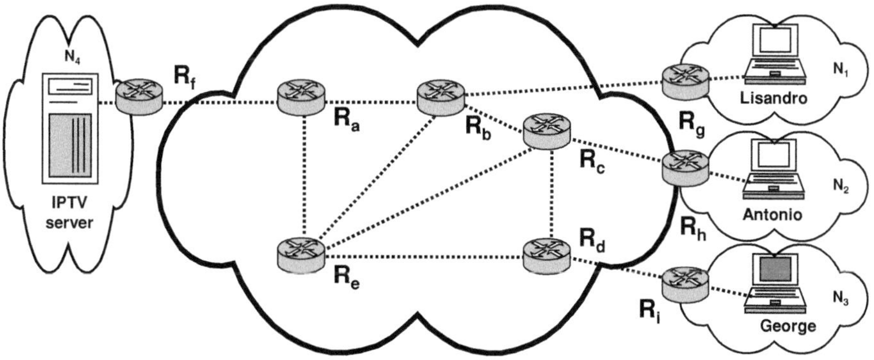

Fig. 2.4 Internetworks

carries its own destination address. The trick is to match this information against the network load in order to determine how to switch the packet at each intermediate node. Ideally, every packet should be routed in such a way as to avoid the situation of Fig. 2.3, where some links get congested whilst others are underutilized.

One must first replace the simple switching devices with proper Internet routers which are able to perform more than just simple switching functions [9]. They have the ability to automatically control the switching process, adapting it to the dynamics of the whole network. Routers connect terminals to networks; though their primary task is to interconnect networks with other networks (Fig. 2.4). In this way, routers provide the fundamental vehicle to transfer data beyond the boundary of individual networks.

Before we can understand how a packet is delivered from end to end, let's follow its maneuvers inside the router (Fig. 2.5). Once a packet reaches the input port of a router, its destiny is to sit in the local memory buffer while the existing population of packets is being digested. Then, the router extracts the packet's network destination address and matches this information against its own routing table. We are now looking into router R_b; it's the turn of one of George's IPTV packets. The network destination address is N_3. The routing table tells us that the packet must be forwarded via output port Out_3 (Fig. 2.6a). We also get an estimation of the distance to George's terminal; though this information is not utilized at this point. The router now ejects the packet via Out_3 and is ready to pick another packet from the input buffer.

Likewise, the scenario of Fig. 2.3, a video call between Lisandro and Antonio, hits the router. In this case, the destination network is N_2. However, the corresponding next-hop address entry in the routing table (Out_3) puts the stream right onto the same output line as George's. Looking at Fig. 2.7a, we can see that the link between R_b and R_c has become the bottleneck. At the same time, there is an alternative path between Lisandro and Antonio ($R_b \rightarrow R_e \rightarrow R_c \rightarrow R_h$) that would bypass the bottleneck. If only we could manipulate R_b's table, changing the second row as in Fig. 2.6b, we would achieve a nicely load-balanced network (Fig. 2.7b).

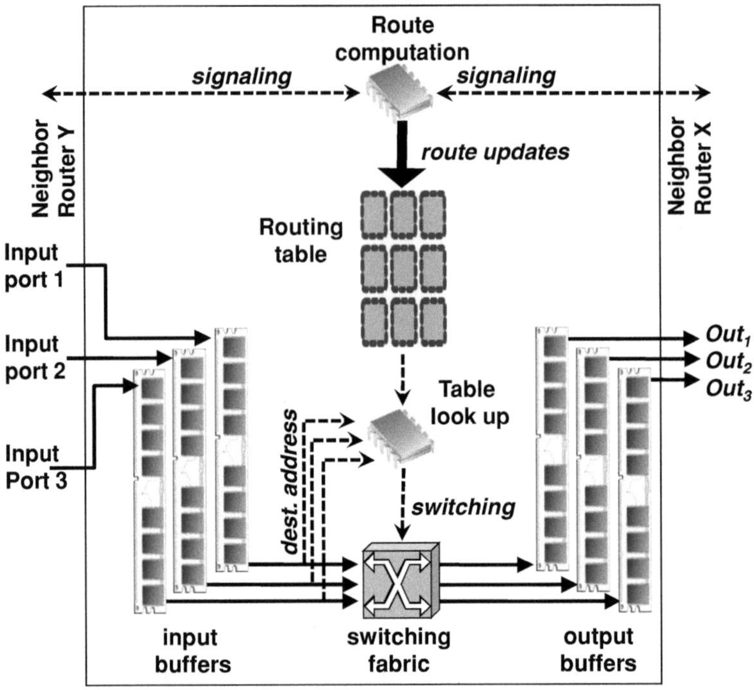

Fig. 2.5 Anatomy of a router

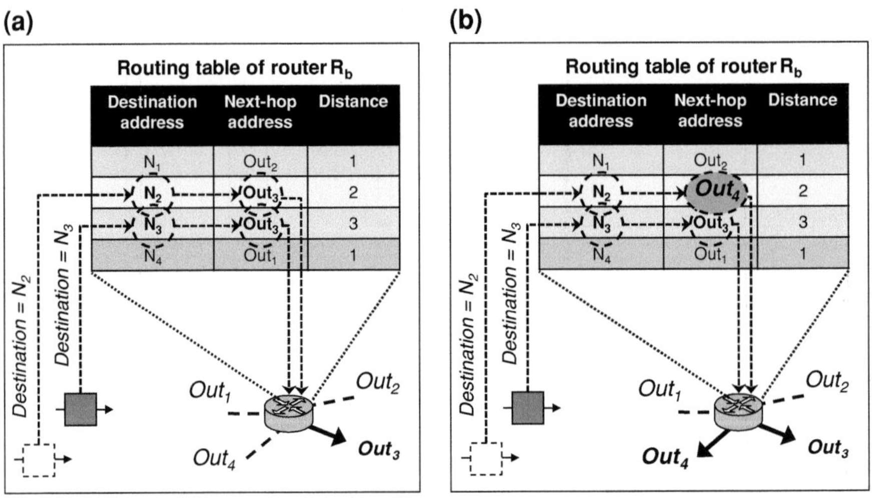

Fig. 2.6 Address matching with the routing table of router R_b. **a** Two different streams are forwarded via the same output interface. **b** A table entry is changed to divert one of the streams

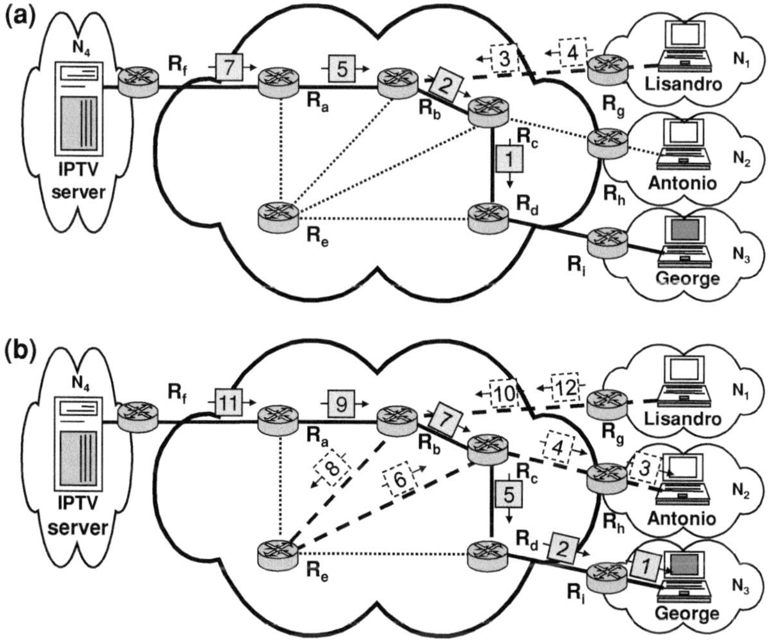

Fig. 2.7 Routing protocols pursue uniform network utilization (load balancing)

The routing tables steer packets throughout the network. Simple manipulations of those tables help to reduce resource contention. Nevertheless, who is in charge of the tables? It is certainly not human operators. The Internet is far too big and dynamic. Each router has full control of its own table. Based on information gathered from other routers (the signaling lines of Fig. 2.5), a router periodically self-optimizes its table, striving for an even distribution of network traffic. We must borrow the skills of a computer scientist to understand how this is realized.

2.5 The Computer Science of a Router

Route computation takes place in the background and in parallel to the store–match–switch–forward process. There are different ways to go about it. Let's start with the Distance Vector (DV) routing protocol (DV is just another name for the routing table) [1]. When a router starts-up, its table is empty. Through the network interfaces, the router can sense its neighbors. Every router periodically sends its DV to the neighbors, which provides information about the neighbors' neighbors. In this way, after a few table exchanges, every router receives information about every other network.

Let's follow the table creation process for the network of Fig. 2.8a. There are four sub-networks (labeled N_1–N_4) and nine routers (labeled A–F). At start-up

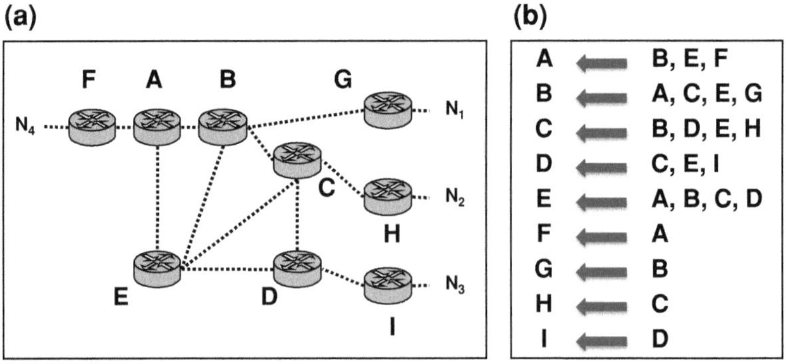

Fig. 2.8 **a** Sample network. **b** Schematic view of table exchanges

Routing table of router A

Dest. address	Next-hop address	Distance
-	-	-
-	-	-
-	-	-
-	-	-

Routing table of router B

Dest. address	Next-hop address	Distance
-	-	-
-	-	-
-	-	-
-	-	-

Routing table of router C

Dest. address	Next-hop address	Distance
-	-	-
-	-	-
-	-	-
-	-	-

Routing table of router D

Dest. address	Next-hop address	Distance
-	-	-
-	-	-
-	-	-
-	-	-

Routing table of router E

Dest. address	Next-hop address	Distance
-	-	-
-	-	-
-	-	-
-	-	-

Routing table of router F

Dest. address	Next-hop address	Distance
-	-	-
-	-	-
-	-	-
4	-	0

Routing table of router G

Dest. address	Next-hop address	Distance
1	-	0
-	-	-
-	-	-
-	-	-

Routing table of router H

Dest. address	Next-hop address	Distance
-	-	-
2	-	0
-	-	-
-	-	-

Routing table of router I

Dest. address	Next-hop address	Distance
-	-	-
-	-	-
3	-	0
-	-	-

Fig. 2.9 Network bootstrapping

time, each router only knows about its own network (Fig. 2.9). The tables of routers A, B, C, D and E are empty because they are not directly attached to any network. At every table exchange, routers receive tables from their neighbors (Fig. 2.8b) and use this information to learn about networks that sit beyond their immediate horizon. The table update process takes place continuously and involves very large tables. Thus, this process must be computationally simple.

Message from G

Dest. address	Distance
1	0
-	-
-	-
-	-

+ 1

Message from G after increment

Dest. address	Next-hop address	Distance
1	G	1
-	-	-
-	-	-
-	-	-

Old table of router B

Dest. address	Next-hop address	Distance
-	-	-
-	-	-
-	-	-
-	-	-

New table of router B

Dest. address	Next-hop address	Distance
1	G	1
-	-	-
-	-	-
-	-	-

Fig. 2.10 An example table update for router B

Figure 2.10 illustrates how router B gets to learn about the existence of network N_1 after just one table exchange. The tables received from A, C and E are empty at this stage. Thus, B can't acquire any new information from those routers right now. Luckily, G has some useful information: it knows about network N_1. The distance between G and N_1 is zero, indicating that the network is directly attached to router G. However, G is one hop away from B and, thus, the distance between B and N_1 will be equal to 1 hop if packets are forwarded via G. In computational terms, this is a very quick algorithm, requiring an incremental function and a comparison between integers (the distance field in the old table and the new one).

Figure 2.11 shows the tables after the first table exchange-update. At this point, not all networks have been discovered (many table entries are still empty). However, the process continues indefinitely. After further iterations, the tables will look as in Fig. 2.12; then, Fig. 2.13; and finally, Fig. 2.14 (full convergence).

Our network is now stable and operational, and will forward packets between networks N_4 and N_3, and (in parallel) between N_1 and N_2, as shown in Fig. 2.15. Compare these new paths with those depicted in Fig. 2.7. Which configuration is more efficient? The two streams now travel through much shorter paths. The "N_4-to-N_3" stream now visits four (rather than five) routers and at least 20% of time is saved. The "N_1 to N_2" stream traverses three (rather than four) routers: a minimum of 25% of time is saved. Remarkably, these new paths are totally disjoint, reducing the probability of contention and packet loss at router B.

Real networks are vulnerable to hardware failure and congestion. DV routing is simple and efficient. Yet, how robust is a DV network? Two examples will help to identify the strength and weaknesses of DV. First, let's break the link between E and D. The causes for this failure may be varied. Perhaps E has a hardware fault or requires re-booting. It might be that heavy traffic is hitting the link, causing congestion. One way or another, router E detects that its neighbor, D, has become

Routing table of router A

Dest. address	Next-hop address	Distance
-	-	-
-	-	-
-	-	-
4	F	1

Routing table of router B

Dest. address	Next-hop address	Distance
1	G	1
-	-	-
-	-	-
-	-	-

Routing table of router C

Dest. address	Next-hop address	Distance
-	-	-
2	H	1
-	-	-
-	-	-

Routing table of router D

Dest. address	Next-hop address	Distance
-	-	-
-	-	-
3	I	1
-	-	-

Routing table of router E

Dest. address	Next-hop address	Distance
-	-	-
-	-	-
-	-	-
-	-	-

Routing table of router F

Dest. address	Next-hop address	Distance
-	-	-
-	-	-
-	-	-
4	-	0

Routing table of router G

Dest. address	Next-hop address	Distance
1	-	0
-	-	-
-	-	-
-	-	-

Routing table of router H

Dest. address	Next-hop address	Distance
-	-	-
2	-	0
-	-	-
-	-	-

Routing table of router I

Dest. address	Next-hop address	Distance
-	-	-
-	-	-
3	-	0
-	-	-

Fig. 2.11 Tables after first exchange

unreachable. For instance, the DV table from D fails to arrive. Router E takes immediate action. It can't afford to forward packets to D, knowing that these will be dropped. Luckily, both B and C have a valid path to N_3. Router C can do better (2-hop distance); therefore, the table update will take place as shown in Fig. 2.16, bringing the system back to stability.

A second example will now show that, unfortunately, DV is not able to recover from various failures. Suppose that the link B $\rightarrow$ G breaks. The subsequence of events is depicted in Fig. 2.17. Before the failure, router B has an entry indicating that N_1 is reachable via G in one hop. However, on the first table exchange that follows the failure, B doesn't hear anything from G (since the link is broken). The distance to G is thus set to infinity (∞). Router C learns that E has a 2-hop route to N_1 and updates accordingly. At the same time, E learns that C has a 2-hop route to N_1 and updates appropriately.

On the second table exchange, this misunderstanding continues. B thinks that the 3-hop route offered by C is valid and incorporates it. C has a 3-hop path relying on E. Therefore, when router E sends the new table showing that its own distance to network 1 has increased, router C has to raise its distance parameter too. The same applies to C. After three exchanges, all distance parameters go up to five. This degenerative process is inexorable and retains the network in an endless loop.

Routing table of router A

Dest. address	Next-hop address	Distance
1	B	2
-	-	-
-	-	-
4	F	1

Routing table of router B

Dest. address	Next-hop address	Distance
1	G	1
2	C	2
-	-	-
4	A	2

Routing table of router C

Dest. address	Next-hop address	Distance
1	B	2
2	H	1
3	D	2
-	-	-

Routing table of router D

Dest. address	Next-hop address	Distance
-	-	-
2	C	2
3	I	1
-	-	-

Routing table of router E

Dest. address	Next-hop address	Distance
1	B	2
2	C	2
3	D	2
4	A	2

Routing table of router F

Dest. address	Next-hop address	Distance
-	-	-
-	-	-
-	-	-
4	-	0

Routing table of router G

Dest. address	Next-hop address	Distance
1	-	0
-	-	-
-	-	-
-	-	-

Routing table of router H

Dest. address	Next-hop address	Distance
-	-	-
2	-	0
-	-	-
-	-	-

Routing table of router I

Dest. address	Next-hop address	Distance
-	-	-
-	-	-
3	-	0
-	-	-

Fig. 2.12 Tables after second exchange

This situation is known as the "count-to-infinity" problem. In fact, all routes to N_1 are invalid but individual routers don't know it. They just keep counting.

DV is a protocol that addresses a global problem (building routes from any network to any other one) with an eminent strategy. Each router plays a part in solving the riddle by carrying out elementary actions (counting and matching). However, DV is the victim of its own simplicity. When networks become large and dynamic, DV is not up to the job. Apart from the count-to-infinity syndrome, can you imagine what happens to those tables as new networks join the global Net? Huge tables will have to be periodically exchanged and reprocessed, with detrimental consequences for both the Net and the routers. This is why the global Net has eventually given up DV for a more controllable approach.

2.6 Simple Math to Stabilize the Net

In the 50s, Dutch mathematician Edsger W. Dijkstra was looking for the best way to travel between different computer circuitries. In those days, computers were far bigger and hotter. Dijkstra's shortest-path algorithm offered a way to convey electricity to all essential components in the most economical fashion. Ironically,

Routing table of router A

Dest. address	Next-hop address	Distance
1	B	2
2	B	3
3	E	3
4	F	1

Routing table of router B

Dest. address	Next-hop address	Distance
1	G	1
2	C	2
3	C	3
4	A	2

Routing table of router C

Dest. address	Next-hop address	Distance
1	B	2
2	H	1
3	D	2
4	B	3

Routing table of router D

Dest. address	Next-hop address	Distance
1	C	3
2	C	2
3	I	1
4	E	3

Routing table of router E

Dest. address	Next-hop address	Distance
1	B	2
2	C	2
3	D	2
4	A	2

Routing table of router F

Dest. address	Next-hop address	Distance
1	A	3
-	-	-
-	-	-
4	-	0

Routing table of router G

Dest. address	Next-hop address	Distance
1	-	0
2	B	3
-	-	-
4	B	3

Routing table of router H

Dest. address	Next-hop address	Distance
1	C	3
2	-	0
3	C	3
-	-	-

Routing table of router I

Dest. address	Next-hop address	Distance
-	-	-
2	D	3
3	-	0
-	-	-

Fig. 2.13 Tables after third exchange

Dijkstra's algorithm has laid the foundations of modern computer networks. We now explore a new technique that overcomes the limitations of DV routing and helps to stabilize dynamic networks.

A big problem in DV is that network-status information propagates slowly and inefficiently. Each router periodically sends its own table to the neighbors. So when something changes around a router (e.g., congestion or failure), only the immediate neighbors are quickly informed. It takes a number of periodic updates before the other routers adapt their tables. In the meantime, the network operates sub-optimally. Also, let's not forget about the count-to-infinity problem.

In Link State routing (LS for brevity), each node talks to all other nodes, ensuring a much faster response [1]. Every router probes the health of its own connections and creates an "update" packet, the Link State Packet (LSP). The LSP contains a list of directly connected neighbors and their distance (or cost). Unlike DV, LS routing broadcasts LSPs, which ensures that any localized network change reaches all other sections immediately. Another advantage is that LSPs are much smaller than DV packets. LSPs only have one entry per network interface. For instance, in Fig. 2.18a, the LSP of router A will have three entries, the LSP of router B will have four entries and so forth.

Routing table of router A

Dest. address	Next-hop address	Distance
1	B	2
2	B	3
3	E	3
4	F	1

Routing table of router B

Dest. address	Next-hop address	Distance
1	G	1
2	C	2
3	C	3
4	A	2

Routing table of router C

Dest. address	Next-hop address	Distance
1	B	2
2	H	1
3	D	2
4	B	3

Routing table of router D

Dest. address	Next-hop address	Distance
1	C	3
2	C	2
3	I	1
4	E	3

Routing table of router E

Dest. address	Next-hop address	Distance
1	B	2
2	C	2
3	D	2
4	A	2

Routing table of router F

Dest. address	Next-hop address	Distance
1	A	3
2	A	4
3	A	4
4	-	0

Routing table of router G

Dest. address	Next-hop address	Distance
1	-	0
2	B	3
3	B	4
4	B	3

Routing table of router H

Dest. address	Next-hop address	Distance
1	C	3
2	-	0
3	C	3
4	C	4

Routing table of router I

Dest. address	Next-hop address	Distance
1	D	4
2	D	3
3		0
4	D	4

Fig. 2.14 Tables converge after forth exchange

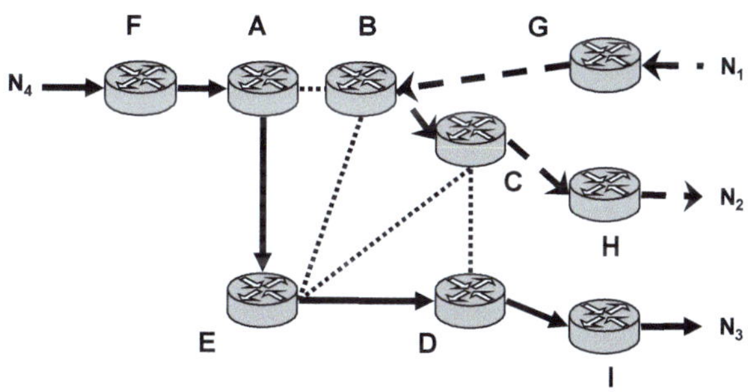

Fig. 2.15 Routing paths between $N_4 \rightarrow N_3$ and $N_1 \rightarrow N_2$

Once a given router has a copy of the LSP from every other router, it is able to compute the best route to each destination thanks to the algorithm that Dijkstra invented in 1956. In his dissertation, he describes the procedure in graph-theoretic terms. If you are not a mathematician, Fig. 2.18 might come in handy. We start off with a simple network (a). It is worth noting that although the topology is the same as in the other examples within this chapter, here, links have a "cost" label. This reflects the distance between adjacent nodes. Though also, it indirectly reveals the level of congestion of the link. As traffic increases on the links, so does the cost. The higher the cost, the more urgently we need to find diversions.

Message from C

Dest. address	Distance
1	2
2	1
3	2
4	3

+ 1

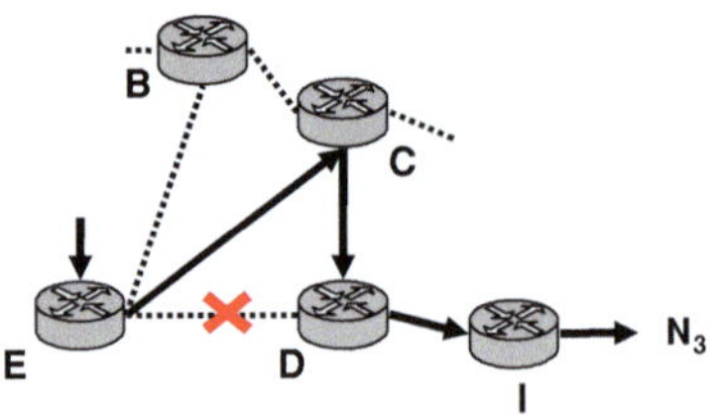

Message from C after increment

Dest. address	Next-hop address	Distance
1	C	3
2	C	2
3	C	3
4	C	4

Old table of router E

Dest. address	Next-hop address	Distance
1	B	2
2	C	2
3	D	∞
4	A	2

New table of router B

Dest. address	Next-hop address	Distance
1	B	2
2	C	2
3	C	3
4	A	2

Fig. 2.16 A problematic link (E → D) is bypassed

After 1 exchange

Routing table of router B

Dest. address	Next-hop address	Distance
1	-	∞
2	C	2
3	C	3
4	A	2

Routing table of router C

Dest. address	Next-hop address	Distance
1	E	3
2	H	1
3	D	2
4	B	3

Routing table of router E

Dest. address	Next-hop address	Distance
1	C	3
2	C	2
3	D	2
4	A	2

After 2 exchanges

Routing table of router B

Dest. address	Next-hop address	Distance
1	C	4
2	C	2
3	C	3
4	A	2

Routing table of router C

Dest. address	Next-hop address	Distance
1	E	4
2	H	1
3	D	2
4	B	3

Routing table of router E

Dest. address	Next-hop address	Distance
1	C	4
2	C	2
3	D	2
4	A	2

After 3 exchanges

Routing table of router B

Dest. address	Next-hop address	Distance
1	C	5
2	C	2
3	C	3
4	A	2

Routing table of router C

Dest. address	Next-hop address	Distance
1	E	5
2	H	1
3	D	2
4	B	3

Routing table of router E

Dest. address	Next-hop address	Distance
1	C	5
2	C	2
3	D	2
4	A	2

Fig. 2.17 B → G breaks down, triggering the "count to infinity" problem

Every router runs the same algorithm. Let's just follow the subsequence of events in router F. The program maintains two separate lists: the list of "tentative" paths and the list of "confirmed" paths. The entries of those lists are formed as {destination, next-hop, cost}.

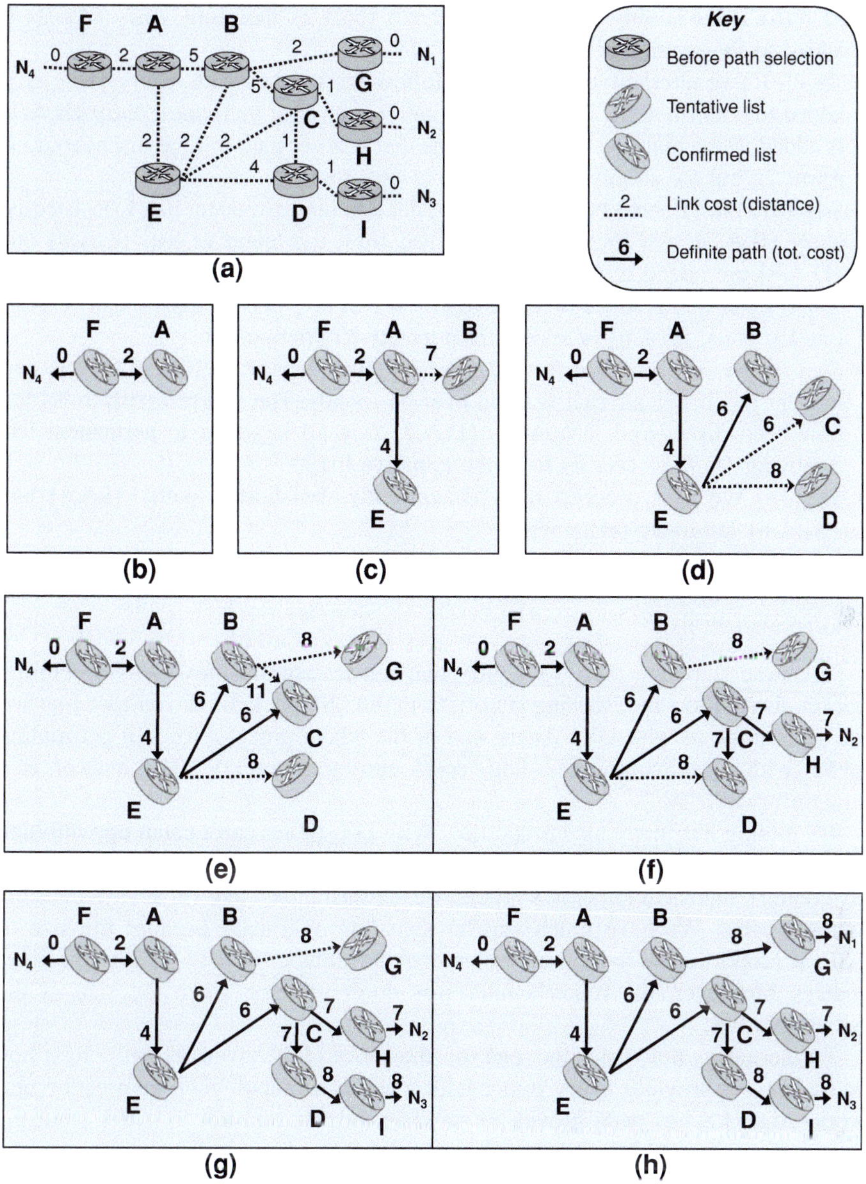

Fig. 2.18 Dijkstra's shortest-path algorithm

- *Status just before Step* (*b*) At the very beginning, the confirmed list contains only the entry {N_4,–,0}: network N_4 is directly attached to router F. The tentative list is empty. Now, it's time to start exploring the shortest path available until a better alternative is found.

- *Step* (*b*) There is only one path to router A (cost 2); therefore, entry {A,–,2} is added to the confirmed list.
- *Step* (*c*) Two alternatives appear; we follow the shortest path. Entry {E,A,4} is added to confirmed list (no better alternatives can ever be found); entry {B,A,7} is added to the tentative list (it might be that a better path will be discovered via router E, but we simply can't be sure at this point).
- *Step* (*d*) Path {B,A,6} (via router E) is added to the permanent list. Consistently, entry {B,A,7} (via router A) is removed from the tentative list. {C,A,6} and {D,A,8} are added to the tentative list.
- *Step* (*e*) {G,A,8} is added to tentative list. {C,A,11} (via B) is longer than {C,A,6} (via E). Thus, {C,A,6} is moved from tentative to permanent.
- *Step* (*f*) We now proceed from C (currently the shortest path). {H,A,7} is permanent (no better alternatives can ever be found). The shortest path to N_2 has now been discovered, {N_2,A,7}. {D,A,7} (via C) is added to permanent list, removing {D,A,8} (via E) from the tentative list.
- *Step* (*g*) We now proceed from D (currently the shortest path). {I,A,8} and {N_3,A,8} are made permanent.
- *Step* (*h*) Having explored all shorter paths, we go back to router G. Entry {G,A,8} is moved from tentative to permanent. {N_1,A,8} is discovered (added to permanent list).

All these steps are there to signify that we are able to inexpensively build a packet distribution tree, starting simply from the LSP packets. Haven't we reached the same point as with DV? At the end of the whole process, we still get routing tables with {destination, next-hop, cost} entries as in DV. The answer is a straightforward "no."

LS routing has many nice properties. With DV, tables can't catch up with high network dynamics. It builds paths based on information that easily becomes stale. By contrast, thanks to Dijkstra's proofs, we are guaranteed that LS always finds the shortest paths. We have much smaller signaling overheads because the size of LSPs is proportionate to the router connectivity (number of network interfaces per router). Moreover, the router's interfaces don't increase when the size of the network grows.

Furthermore, this isn't the end of the story. DV gives us only next-hop addresses. LS provides every router with a complete topological map of the network. Also, LS has been proven to stabilize quickly and not to suffer from the count-to-infinity problem. It's not in the style of this book to bombard the reader with formulas, theorems or proofs. Nevertheless, here is a little scenario that will hopefully convince you.

Figure 2.19a again proposes the same topology. Though now, we pump packets into the network. A number of users sitting in networks N_1, N_2 and N_3 start streaming videos from N_4. The optimal distribution tree calculated before (Fig. 2.18h) pushes all streams through A $\rightarrow$ E. This is bottleneck! E $\rightarrow$ C is another one!

If you are wondering whether Dijkstra actually did a good job, think again. He did a fantastic one! His distribution tree was perfect, considering the network

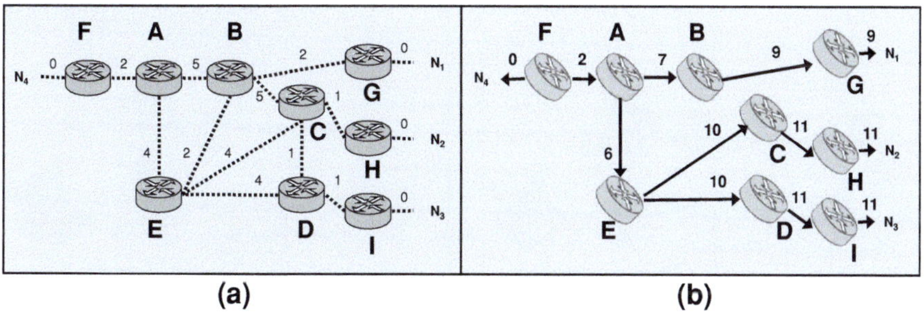

Fig. 2.19 Traffic equalization: **a** network status; **b** distribution tree of router F

utilization status captured at the time when router F made its calculations. Though now, the network load is changing. Chances are that we need new diversions. The beauty of LS is that it will actually be able to equalize the network. As soon as traffic piles up in those two bottlenecks, routers A, E and C will immediately sense higher latencies in their local links. The costs associated with links A → E and E → C will go up. Dijkstra will kick in again with new paths. Assuming that the link cost of our bottlenecks goes up from two to four, we end up with the new traffic distribution tree depicted in Fig. 2.19b. Again, we have a well-balanced network.

2.7 Life of a Commuter

So far, we have seen how fixed networks remain stable, despite failures and traffic dynamics. How does the network cope with travelers? Most networks and hotspots use automatic configuration methods to help connecting new terminals. An example is the Dynamic Host Configuration Protocol (DHCP) that assigns a unique address to your laptop when you are sitting in the airport lounge. Your IP address includes the address of the network you are physically attached to and makes you reachable from any router.

For example, you decide to take a walk and get a coffee somewhere in the airport in order to better digest your full email inbox. By doing so, you lose connectivity but, nicely enough, another Internet provider covers the café. Different provider; different network identifier; different IP address. You gain a new identity in the Net and can resume your email gymnastics.

Let's rewind the scenario. You've been promoted from "workaholic" to "TV addict." After authenticating yourself with an IPTV provider, filling in all your personal details and making a transaction with your precious credit card, you have started watching your preferred series. However, you are still desperate for a coffee. This time though, you find that, following the network handover, the session is broken. You'll have to start the authentication process all over again! What's different now is that the IPTV session includes authentication,

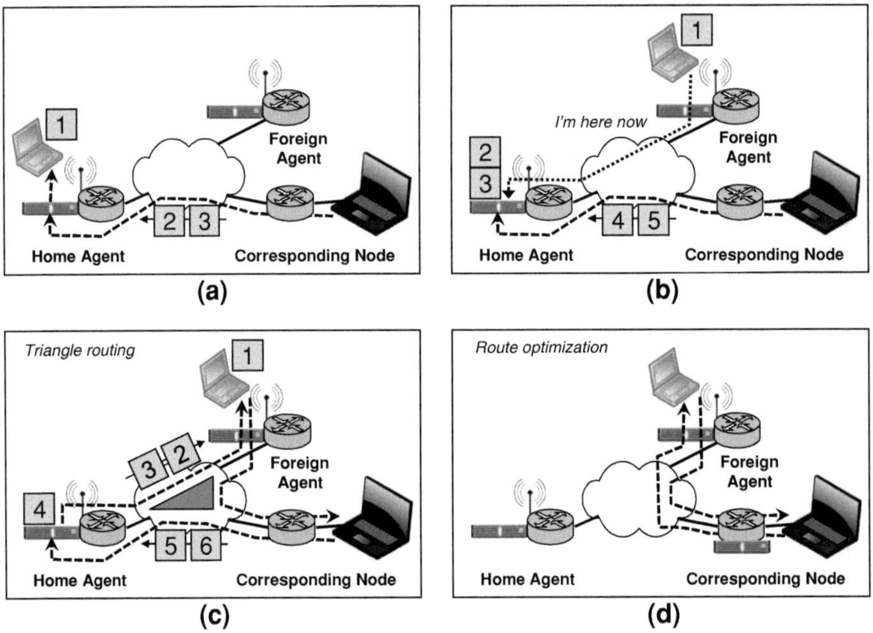

Fig. 2.20 Routing with mobile terminals

authorization and status information that is tightly linked with the IP address. Thus, when you handover, you lose it all.

New technologies are emerging to keep sessions alive during roaming. For instance, cellular networks have their own specific approach. Unfortunately, that would bring us far beyond the horizon of this book. We are looking for a more universal (non-technology-specific) solution, allowing terminals to retain their IP address, regardless of their network access point [10].

A possible stratagem is depicted in Fig. 2.20. In the first sketch (Fig. 2.20a), a stream of packets goes from the corresponding node to the mobile terminal. Notice that the edge router of the mobile terminal is re-named as "home agent." This is like an ordinary router with just the extra ability to act as a proxy agent for the mobile terminal. In Fig. 2.20b, we see the counterpart of the home agent, the "foreign agent." As soon as the terminal regains connectivity, it lets his home agent know about the new foreign agent address. The home agent now knows how to reach the terminal and can relay the stream (Fig. 2.20c). The reverse path doesn't have to follow this procedure. The terminal can reach the corresponding node directly. We notice here a routing anomaly, referred to as the "triangle routing." But it's relatively easy to resolve this sub-optimal condition. The terminal knows the address of the corresponding node and can let it know about the foreign agent (Fig. 2.20d). The corresponding node can create its own tunnel to the foreign agent, avoiding the detour.

You've just learnt the essence of Mobile IP [10, 11], an ongoing effort to try and add mobility to the Internet without modifying the core routers. In practice, minor modifications are required only on the edge routers in order to realize the "agent" functionality. With Mobile IP, the routing tables of the core routers need not be modified after you change the attachment network.

2.8 The Three Fundamental Principles

As complex as it might seem, the Internet performs its task of keeping billions of terminals connected thanks to just a few elementary operations: routing, buffering and forwarding. The Internet architects managed to create a "generative" framework, an open network imposing very little constraints to the application developer. The outcomes of this approach are clearly visible: the Web continues to generate innovation and has now become the very backbone of our social and business lives.

Network technologies evolve relentlessly in speed, mobility and automation. Thanks to the unifying power of the Internet Protocol (IP), all sorts of networks have formed an intricate web of connections. Packets are transported across diverse chunks made of copper, fibers and radio waves. However, the Internet is not the only and not necessarily the best way to keep billions of things connected. Many scientists have started to question the ability of the Internet to sustain its current growth rate. Terminals are becoming far more volatile than they were just a couple of years ago. Not only cellular phones, but also, sensors, security cameras, televisions and all sorts of appliances are all joining the Net. Yet, this unrelenting revolution is confined to the edge of the network. Except for incremental bandwidth upgrades, the network core hasn't changed for decades.

Engineers have a motto: "if it works, don't fix it!" Unfortunately, the Net is cracking under the weight of the "ubiquitous" user, file sharing, IPTV and the who-knows-what's-coming-next pervasive application. Because of this, more often than ever, the buzzword "Future Internet" appears higher up on the agendas of politicians, national funding bodies and academic circles. After all, following the edge-network revolution, we might soon witness the core-network revolution. Though, until somebody comes up with a bright new idea (a fundamental paradigm shift), we should attempt to make the most out of what has already been researched.

To pursue this intention, in the following chapters, we'll revisit some clever networking mechanisms that researchers worldwide have been studying for a number of years. We are not interested in all the technicalities of the myriad of proposed protocols. It's rather more constructive to pull out the key concepts and ideas that lay behind them.

Before we start looking for new networks, it's instructive to reason just a bit more. Thus, we start this journey of "concept extraction" from the Internet itself.

What are the lessons learned from the Internet phenomenon? The Net won the natural selection battle against many other proprietary networks that were far better engineered. The Internet resulted from an extreme process of simplification. Yet, it has become an indispensable tool. Hence, by observing the Net, we have the unique opportunity to discover the very essential elements of networking—the fabric that makes any network. We believe that the intriguing digital ecosystem that we see today is based merely on three fundamental principles:

1. Connect
2. Discover
3. Stay connected

Before you can enjoy the wonders of the Net, what's the very first thing you ought to do? Plug-in your network cable or get attached to a wireless hotspot. In either case, there is some sort of authentication procedure through which the network checks your credentials. Only then are you assigned an IP address; you now have a Net identity; you can reach and can be reached by the digital mass; you are connected!

The process by which one obtains a unique IP address was cumbersome up to just a few years ago. Thanks to a bunch of discovery protocols, auto-configuration, DHCP, Mobile IP and other acronyms most people can survive without, the "connect" bit of internetworking has become totally seamless. Do I really need to appreciate the difference between plugging in my power adaptor and plugging into the Net? From a person's viewpoint, both are just plugging exercises. We harness energy through the power plug and information through the Net plug. What happens beyond the plug is not of everybody's concern.

In the not-so-far future, we may witness a sophistication of the "connect" step. Security and privacy requirements and regulations are gaining importance; people and organizations are increasingly wary of identity theft, spam, spyware, viruses, phishing and the lot. We all want to gain access to the digital information society. Though, increasingly, not all of us want to do it at "any" cost. We want to know what we are connecting to; whether it can be trusted with our credit card details; how far it regards our privacy. The Internet can't do any of this because it's just a packet-forwarding machine. It's not aware of *who* is doing *what*. Further along in this book, there will be ample opportunity to explore better "connect" mechanisms and to get inspiration from the milestones achieved by the P2P community.

Let's now explore the second principle of networking: "discover." You joined the Net for a reason; to do something. You'll need to find the necessary resources before you can carry out any task. You've read earlier in this chapter how paths from one point to another are discovered via the forwarding tables maintained by the routers. The routing algorithms currently in use are as simple as they can get. So far, they have done their job; but the future doesn't look too bright.

It's again that mixture of time-constrained and data-intensive applications that rows against the tide. IP doesn't differentiate among the different flavors of data flows. It finds "shortest" paths that are totally data agnostic. Packets carrying an email flow travel in the same "class" as their fellows who are in charge of a video call. Equality is a nice notion, though only if we are all pulling the same weight. If a few email-packets get delayed by one second, nobody is affected. But if the same happens to our audio-packets, the application will inexorably break. The cost metric must be differentiated if we want to make the most of our limited resources. In many situations, it might be more efficient to treat flows in different ways, but the Internet only adopts one set of tables for all packets. The Net copes with increasingly tight time-constraints only in one way: over-provisioning of network resources. For how long shall we be able to afford this luxury? Once more, the Net sacrifices efficiency in the name of simplicity.

Now, let's return to the three principles. We are connected and are able to find meaningful paths. This is still not enough! The network is dynamic; it's alive; it suffers failures; it gets congested; it's vulnerable to external strain. Nevertheless, the Net is able to "absorb" a good extent of variability in order to keep us connected. However, if your Skype session has just dropped unexpectedly, this is because the network has failed to adjust. Remember, applications are tolerant to network impairment only to a limited extent. Some applications are better than others, but they all need to chew packets within specific deadlines.

Given these premises, where are we heading to? The Internet was conceived with some key requirements that still hold today:

- *simplicity* keep complexity to the minimum;
- *general-purpose fabric* purposely not optimized for any specific application to give space to creativity;
- *scalability* thanks to its distributed architecture, it can be easily upgraded to cater for more users;
- *resilience* no single-point of failure; and multiple paths make it robust against localized attacks;
- *self-control* no need for any centralized management platform.

Unfortunately, not all the original design assumptions are still valid. Something very fundamental has changed which makes it more and more difficult to keep people connected. Entirely new dimensions have come into the picture:

- *extreme mobility* with all sort of wireless and cellular technologies at the network edge;
- *disruptive applications* with sharing of files and streams at an unprecedented rate;
- *content producers* anybody, not only the media companies, can generate multimedia content and inject it into the Net.

In the following chapters, we are going to see other kinds of networks empowered by more sophisticated mechanisms than IP. Yet, the three principles will stay the same. Together, we'll explore ideas that can make the Internet even more pervasive and generative than it is today.

References

1. Peterson LL, Davie BS (2007) Computer networks: a systems approach, 4th edn. Morgan Kaufmann, San Francisco
2. Stallings W (2010) Data and computer communications. Prentice Hall, Upper Saddle River
3. Tanenbaum AS (2002) Computer networks. Prentice Hall, Upper Saddle River
4. Agboma F, Liotta A (2009) QoE in pervasive telecommunication systems. Pervasive computing innovations in intelligent multimedia and applications. Springer, London, pp 365–382
5. Dollimore J, Kindberg T, Coulouris G (2005) Distributed systems: concepts and design. Addison Wesley, Reading
6. Tutsch D (2010) Performance analysis of network architectures. Springer, Berlin
7. Zink M (2005) Scalable video on demand: adaptive internet-based distribution. Wiley, Chichester
8. Firestone S, Ramalingam T, Fry S (2007) Voice and video conferencing fundamentals. Cisco Press, Indianapolis
9. Halabi S (2000) Internet routing architectures. Cisco Press, Indianapolis
10. Soliman H (2004) Mobile IPv6: mobility in a wireless Internet. Addison-Wesley Professional, Boston
11. Loshin P (2003) IPv6: theory, protocol, and practice, 2nd edn. Morgan Kaufmann, San Francisco

Chapter 3
Six Problems for the Service Provider

Abstract The Internet is far from being perfect, and thus we are bound to see many remarkable changes in the near future. In examining what the Net can and cannot do today, what are the top six reasons to upgrade it? The Net is not sufficiently *ubiquitous*, *reactive*, *proactive*, *information-driven*, *distribution-efficient* and *searchable*. This chapter introduces these widely recognized issues, paving the way for the solutions presented in Chap. 4 and thereafter.

The Internet only just *works*
Mark Handley, Professor at UCL, UK

3.1 The Net has Ossified

There are an infinite number of reasons to advocate a complete overhaul of the Internet, though only one is undisputable: the Net has ossified. The general-purpose connectivity machine conceived in the 1970s has now become too vast to afford any significant alteration. Today, almost two billion people use the Net. Each terminal practically has a distinct configuration if we consider the variety of terminals available off-the-shelf. This uniqueness is further defined by the range of software running on each terminal, including operating systems, firewalls, anti-virus and personal applications. We all enjoy customizing computers and phones, though many end up misconfiguring and destabilizing their own systems. And then, there are several varieties of viruses, Trojan horses, spyware and the lot. All sorts of stable as well as unstable machines are attached to the Net.

Thus, one would expect the Net to be periodically upgraded to cater for new terminals, usage patterns and threats. Yet, all the attempts made in the last 15 years to modify fundamental network mechanisms have failed. In principle, the Net

A. Liotta and G. Exarchakos, *Networks for Pervasive Services*,
Lecture Notes in Electrical Engineering, 92, DOI: 10.1007/978-94-007-1473-1_3,

could easily be modified. IP is an extensible protocol. It allows optional instructions to be embedded with packets for special treatment. On the other hand, processing extra instructions considerably slows down the router. Furthermore, many routers actually filter out anything but the simplest IP packets in order to prevent denial-of-service attacks. Thus, the use of protocol extension mechanisms has been lost over the years. IPv6, the latest Internet routing protocol, has been trying to resuscitate IP-option fields; but even the simple transition from IPv4 has been hampered by endless difficulties.

Before getting ossified, the Net was indeed able to upgrade. In 1980, Distance Vector routing was replaced by the more dynamic Link State routing approach. In 1982, the Domain Name System (DNS) introduced a distributed mechanism to uniquely and dynamically name Internet systems. This step became essential as soon as it was clear that the Internet started growing exponentially. Just a year later, another architectural change took place: the mechanics of "reliable transport" (the Transmission Control Protocol or TCP) was separated from the actual "packet delivery" protocol (IP) [1].

Initially, TCP allowed for the identification of missing packets and triggered re-transmission. It was not able to handle network congestion. Hence, in 1988, TCP was extended with mechanisms that could control the rate of transmission of packet flows. Other improvements followed, that is, until 1993, when the Internet reached over two million hosts. At that point, the routing tables were becoming far too large; routers were struggling due to limited computational and memory capacity. Thus, Classless Inter-Domain Routing (CIDR) was introduced, revolutionizing the Internet addressing architecture [2].

The year 1993 is an important turning point for the Net. Mosaic, the very first Internet Browser was released, giving birth to the World Wide Web [3]. However, this is also the last time we would see any significant upgrade into the "core" network. After CIDR, all other attempts to modify the Net failed. The Net was too big and too complex. Any further innovation started happening at its edge, rather than in the core. The ossification process had started, inexorably.

Despite several attempts to accelerate its engine, the Net still works on a single gear: it moves packets "just about" fast enough. However, different applications (video, voice, gaming, etc.) operate over different time constraints; thus, new gears are needed. Researchers have put remarkable effort into trying to find practical ways to migrate away from the "best-effort" nature of the Internet. The Integrated Services (IntServ) framework was the first to design a new gearbox for the Net [4]. IntServ was even standardized in 1994, though following the hype, it became clear that a fine-grained approach operating on each individual data-flow would not work on a large scale.

The Differentiated Services (DiffServ) architecture followed another path [4]. To pursue scalability, it standardized a coarse-grained mechanism to differentiate among different classes of traffic. DiffServ was also standardized (in 1998), but after a dozen years, it is hardly ubiquitous. Only some ISPs adopt DiffServ and there doesn't seem to be a sufficient amount of pressure to cause global-scale deployment.

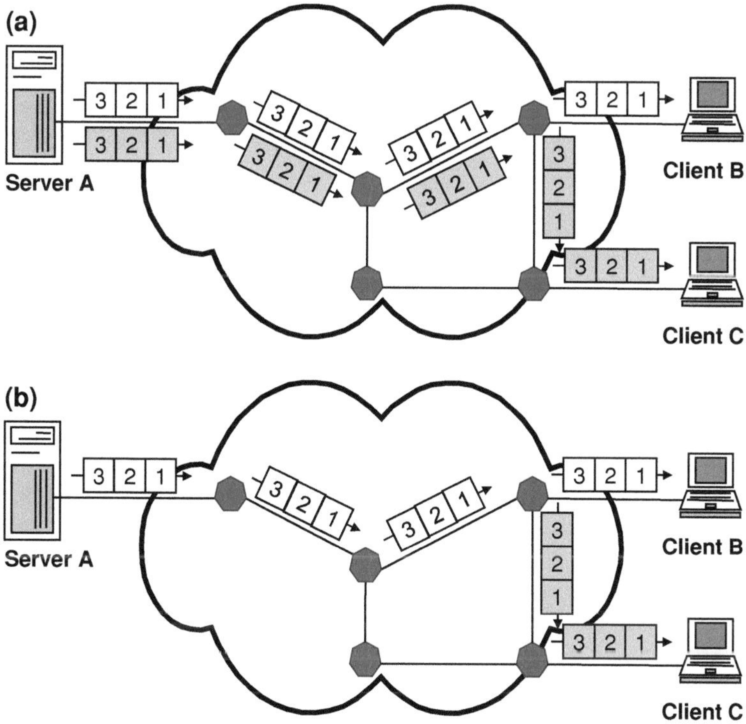

Fig. 3.1 Stream distribution mechanisms in an ordinary wireline IP network. **a** Multiple unicast transmission; **b** multicasting transmission

Other attempts to add new dimensions to the Net were made in the 1990s. Given the prominent trend to distribute audio and video in packetized forms rather than over the more conventional radio and television signals, the Net was missing an essential mechanism: "broadcasting." The Net is good at transmitting packets from any point to another. This is good enough for server-to-client transmission or person-to-person communication. However, to deliver the same stream to a large number of people simultaneously, the Net can only rely on "multiple unicast" (Fig. 3.1a). This is a fairly rudimental distribution mechanism, unsuited to applications such as IPTV. Imagine one thousand people connected on the same channel. One thousand copies of exactly the same stream are generated by the server and injected into the Net.

It isn't necessary to go to great lengths to find a better distribution mechanism. In the early 1990s, researchers came up with a wealth of mechanisms under the banner of "multicast." In essence, when more people want to get the same stream, they register with a special "multicast" address. The Net then builds a multicast distribution tree, as exemplified in Fig. 3.1b. In this way, the Net can make sure that no link ever carries duplicate packets. Native network support to "multicast transmission" was the great promise of the 1990s. Unfortunately, "multicast" was never able to make it to the global Internet.

"Quality of Service" and "Multicast Distribution" are just two of the many examples in which the Net failed to upgrade, pushing the problem towards its edges. Mobility support is another unsolved issue. Today, the majority of terminals are "mobile;" though the Net's routing protocols are not geared for mobility. Even Mobile IP, described in Chap. 2, went successfully from "research" to "standard," but is still struggling with deployment issues [5].

The Net's stagnation is evident in many other sectors: security, privacy and congestion control. Being totally unaware of what the application is doing, the Net is exposed to great risks [6]. In 2009, 81% of emails were spam, accounting for about 73 trillion emails; yet, the Net is disarmed against spam. Denial of service attacks can bring a large corporation to a halt; however, protection against such attacks relies largely on human intervention.

The Internet was never meant to be 100% reliable, efficient, and effective. Its greatest strengths were "malleability" (to adapt easily to new requirements) and "generativity" (to spark new technological advances). For one reason or another, both of these features have gradually been delegated to the periphery of the network. It is clear that the most extraordinary innovation has taken place since 1993. Ironically, the new generation of network mechanisms has found more fertile grounds outside of the network core. A new breed of applications, the "virtual networks," [7] allows the realization of new network mechanisms on top of the physical network.[1] Paradoxically, it is easier to test and deploy new networks in this way, leaving the ossified network unchanged.

However, virtual networks can only carry "virtual" packets. The real load eventually ends up being transported by the underlying "physical" network. The ultimate Net must be able to incorporate and natively support the innovative mechanisms that are currently operating on virtual networks. That is why, from the next chapter onwards, we shall revisit network mechanisms that have matured beyond the Internet core, with the view that these might form the basis for the next-generation Internet.

Before embarking on this task, it's worth spending a few more pages simply reflecting on more reasons to upgrade the Net because today, *the Internet only just works* [8].

3.2 Problem 1: Not Truly Ubiquitous

How "ubiquitous" is the Internet? One could get a first insight by looking at the statistics of Internet penetration expressed as a percentage of population. As of January 1, 2010, an average 26.6% of the global population had Internet access (source: internetworldstats.com). North America leads with a 76.2% penetration, while in Africa, only 8.7% of the population is in the Net.

---

[1] Virtual Networks will be explored in greater details in Chap. 7.

However, the Internet penetration stats only provide a fairly restricted view. They account for the number of people who have some form of subscription with an ISP, for example, cable, DSL, Fiber and so forth. However, ISPs concentrate their infrastructure within the more densely populated areas. A truly ubiquitous network should give connectivity everywhere, or, at least, in any place that is reachable by a human being. On the contrary, the greatest portion of the globe is not on the Net. We can only connect where a physical infrastructure has been put into place.

While the Internet penetration keeps increasing, will the Net ever become "truly" ubiquitous? Will it ever be possible to extend the reach of the Internet beyond the confines of ISPs and Network Operators?

Economic considerations suggest that, in its current form, the Internet cannot ramify ubiquitously. Despite its apparent simplicity, the Internet has grown into an entanglement of complex hardware and software entities. Widespread coverage requires investments to deploy and manage network equipment. Backbone networks are essential in order to keep cities and countries connected; however, do we really need dedicated network devices to increase the capillarity of the Net?

Ironically, network access is often needed where there is no infrastructure. A communication network might help coordinate the efforts of a team of engineers whose task is to actually build a network. Communication is also vital for disaster management, i.e., when a catastrophic earthquake takes place. Yet, in this situation, the infrastructure if often affected by power or hardware failures.

The Internet relies on too many devices and services for its operation. Communication lines, optical fibers, repeaters, transmitters, switches and routers have to be powered up. Protocol stacks must be coherent among each other. Name and address-resolution servers have to be reachable. Then, inside the home, wireless modems and computers have to be consistently configured. Despite all of this complexity, we are still far from a geographically ubiquitous Net.

Going back to the very essence of the Internet, that is, to its original simplicity, it is evident that we don't actually need a sophisticated infrastructure to build a network. Researchers have devised simple techniques which allow individual computers to cluster together and form networks. The "one laptop per child" program developed an inexpensive computer which allowed pupils to connect with each other in places where no other Internet infrastructure was available.

Capillary networks, spontaneous connectivity and on-demand networks can be achieved beyond the strict boundaries of the Internet. Most of today's Internet terminals, including mobile phones, have the capability to perform the same functions of routers. Thus, nowadays, as more and more people carry Internet-enabled devices, it is possible that these form networks which can relay packets just as any ordinary infrastructure-based network would do. Such networks, known as Mobile Ad Hoc Networks (MANETs) [9], hold the magical ingredients of "ubiquity." The next-generation Net will have to incorporate mechanisms to enhance its reach even where no infrastructure has been put in place. To better understand the enchanting world of infrastructure less networks, we shall further explore MANETs in Chap. 4.

3.3 Problem 2: The Unresponsive Net

The Internet routing protocols were designed at a time when networks were neither mobile (wireless) nor fast (optical). Networks were static, that is, their topology didn't change frequently. User terminals were static—they were pre-configured to operate on specific network access points. As the Net started growing, it became clear that even static networks required periodic route re-computation. In fact, Link State routing replaced Distance Vector routing in 1980.

Since then, anything but the very core network has turned into "mobility" mode. There are now more mobile terminals than desktops. People travel and commute more than ever before. Files and streams appear and disappear from various locations in peer-to-peer systems. People Tweet, Skype, email and chat from their mobile phone.

Despite all, the Net can still handle mobility to a certain extent thanks to a plethora of "patches," i.e., stratagems that insulate the core network from the peripheral mobility. However, the trend is for even greater degrees of mobility. It is not only people who access the Internet on-the-move. Networks move as well. Cars, trains and plains are "moving" networks. MANETs are formed by mobile terminals which compute routing tables and relay packets on behalf of other terminals. These, too, are mobile networks characterized by an extremely dynamic topology.

The Net is formed by a relatively slow-changing backbone topology, though even the backbone experiences the side effects of mobility. The traffic going through it is increasingly dynamic. As congestion afflicts a certain network segment, traffic will have to be re-routed via alternative avenues in order to prevent data loss.

High levels of mobility and traffic dynamics require a continuous and prompt recalculation of routes. However, the control engine of the Net is not sufficiently "reactive" for this challenge. The Net tries to maintain routes for everybody in the same way, regardless of whether one moves or remains static. The overall result is that routes exist and are recalculated even when they lead to idle terminals. This "brute-force" approach drains resources from the Net unnecessarily, disadvantaging those nodes that are more dynamic and would require greater attention.

This is yet another symptom of the Net's ossification process, making it difficult to implement diversified mechanisms. How should routing evolve in order to become more reactive? What are better strategies for the Net to keep up with the myriad of mobile "things" that inject and receive packets? In Chap. 5, we shall revisit some tricks that can make routing more reactive and responsive.

3.4 Problem 3: Too Much, Too Stale Signaling

In large systems, it is crucial to make sure that the different subsystems stay in synch in order to operate harmonically. To maintain network paths, each element informs the others by sending signaling control packets. For instance, recall from

Chap. 2 that routers based on the Link State (LS) protocol broadcast LS Packets to inform everybody about the status of the local links. Then, the Dijkstra algorithm reconstructs the topology and builds the optimal routes.

As networks grow, so does the overall signaling traffic. The control packets become larger, travel longer distances and take more time to reach the furthest network elements and be processed. High dynamics make things worse by triggering new signaling packets as soon as something changes. Thus, mere "reactivity" does not always meet the needs of dynamic systems and is, in fact, often the actual source of "instability."

The Internet is not able to adapt its signaling intensity to the network topological characteristics. Highly dense networks require completely different control actions than sparse networks. Mobile nodes demand different trade-offs than static ones. Thus, by using just one calibration strategy, the Internet fails to tailor its control action to the variety of situations that the Net is meant to sustain. Those very same control signals which carry vital "network status" data often overwhelm the Net itself with outdated information.

In principle, routing protocols could be adjusted to the network dynamics. For instance, the frequency of the control packets could be increased or decreased to cater to more or less dynamic conditions, respectively. Different parameters can be used to account for congestion such as hop-count, point-to-point latency or packet loss. However, in practice, these features are only setup at installation time. A more effective approach would be to "proactively" tune the signaling protocols in order to minimize its intrusiveness and maximize its efficacy. Yet, in the name of simplicity, this avenue is not pursued.

The Net is not "proactive." It makes no attempt to anticipate congestion. It cannot recognize bad trends which rapidly escalate into severe congestion. It does not know how to identify typical or repetitive communication patterns. It does not correlate even the simplest events. It cannot learn from the past; nor can it predict, prevent or cure.

The Net can only react (slowly) "after" the effects of congestion become evident. A primary notion in science, engineering and medicine is that "preventing is better than curing." Furthermore, trying to "cure" the root cause of a repetitive problem is certainly better than merely "patching" each and every instance of the same problem. By not treasuring the "prevent or cure rather than patch" principle, the Net is doomed to suboptimality. Its responsiveness is intrinsically circumscribed.

In Chap. 6, we shall see some mechanisms that can reduce unnecessary signaling overheads and, at the same time, exhibit "proactivity" in very dynamic networks. The examples presented therein serve the purpose of demonstrating the potential of "proactive" networks. As the time-scales demanded by the latest real-time services get smaller and smaller, new proactive mechanisms will be required. Reactive and proactive behaviors will have to be harmonized to the extent necessary in order to provide the highest levels of responsiveness and stability.

3.5 Problem 4: Lack of Parallelism

The Net is meant to transport data from point A to B, for instance, from a server A to a Client B. If the path between A and B gets congested, the Net must come up with an alternative path. We have already discussed how this process of discovering new paths and switching over data flows creates problems when the system dynamics increases. Is there anything that can be done to tackle congestion from a different angle? [10].

The Net tries to maintain the shortest path between A and B. Thus, if the distance between A and B computed on the active path tends to increase, the Net assumes that there is congestion. So if any of the links belonging to the A-to-B path becomes a bottleneck, the whole path has to be replaced with a better one.

There is a fundamental flaw with this approach: link congestion is not the only reason why B might not be receiving packets on time. There are many occasions when there is something wrong with the transmitting node (A) such as:

1. Node A might have started moving fast, incurring a higher degree of packet loss;
2. Node A might have received an increased number of requests, reducing its ability to respond promptly;
3. Node A might be subject to a denial-of-service attack which impedes any response at all.

When the transmitting node becomes unfit for this purpose, all efforts made by the Net to discover better paths are misplaced. The Net tries to keep A and B connected (Fig. 3.2a), but it is missing an important point: the key goal is to make sure that B "receives" A's data, not that the data is necessarily coming "from" A [11]. Hence, a better strategy would be to look for a new sender C (Fig. 3.2b), rather than trying to fix the A-to-B path [12].

The Net lacks the necessary mechanisms to realize this level of parallelism. It is designed to multiplex multiple, point-to-point communications. However, it is not able to support parallel transmissions. In the Internet of today, it is common to find that the same information is duplicated in multiple servers. With P2P applications, the same information is actually duplicated on multiple users' terminals [13]. Thus, ideally, we would want the Net to be able to identify a number of transmitters that can simultaneously contribute different "chunks" of the stream (or file) to B in order to minimize the overall transmission time (Fig. 3.3) [14].

To better understand how the future Net could support parallel data transport, Chap. 7 explores concepts and mechanisms developed in the context of P2P networking [7]. The way in which those ideas might 1 day be realized into the core network rather than via P2P applications is still the subject of controversial research.

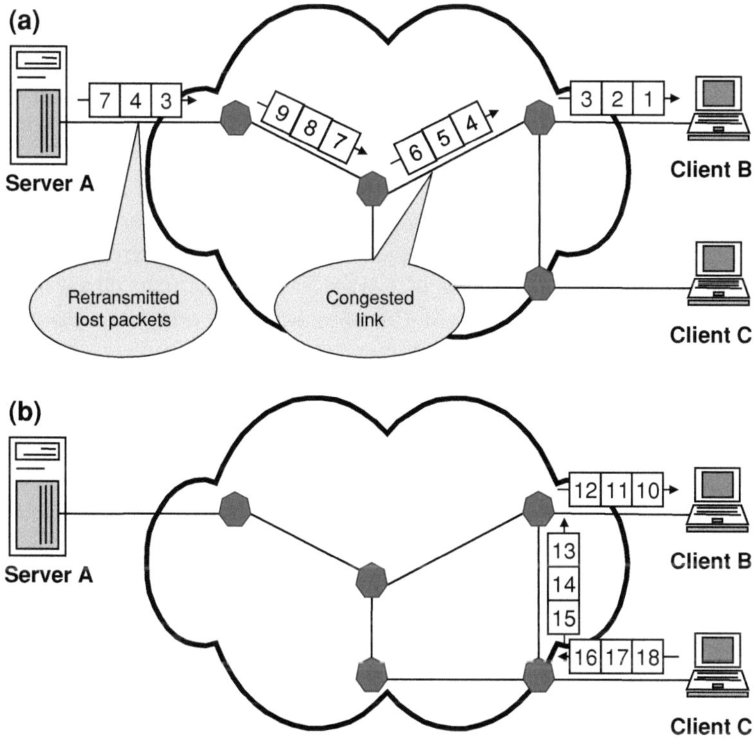

Fig. 3.2 **a** The Net tries to keep nodes A and B connected: lost packets are retransmitted. **b** A better strategy would be to find a different transmitter (node C) that gives a better transmission path

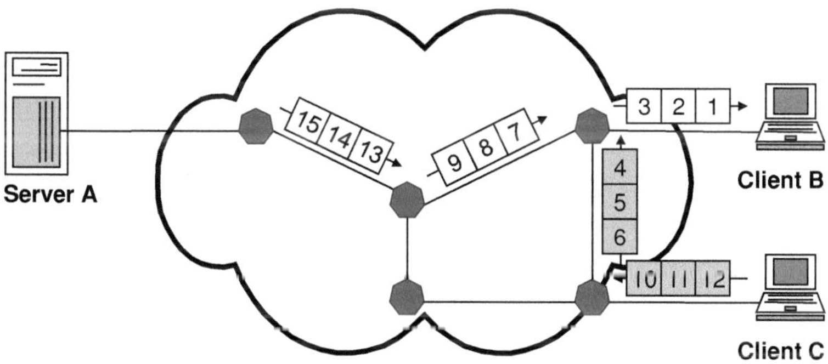

Fig. 3.3 Chunk-based parallel streaming

3.6 Problem 5: Data Agnosticism

Lack of parallelism is quite compatible, so to speak, with another problem: the Net treats packets merely as "raw" entities, that is, a collection of bits sharing their destiny of going from A to B. For the sake of "simplicity" and "fairness," routers ignore the fact that each packet has its own "purpose" which makes the packet "more" or "less" critical. A packet may be part of a video frame. However, not all frames are of equal importance. For instance, in certain video encoding schemes, some frames (the Predicted frames or P-frames) are built upon others (the Intra-coded frames or I-frames). Thus, if P-frames become corrupted (due to packet loss), the overall effect on the perceived quality is relatively minor. However, when I-frames get corrupted, this affects several P-frames, causing a more dramatic quality degradation.

Routers do not want to know whether a packet belongs to an I-frame, a P-frame, an audio sequence or a web-browsing session. In fact, when traffic gets too heavy, routers discard packets somewhat randomly. This causes some transmissions to stall, even under a fairly low packet loss ratio.

The Net is "data agnostic." It does not recognize that packets belong to data-flows and that these are part of a specific user's data sessions. Because of that, we miss the opportunity to deploy more efficient routing algorithms.

One way to make the Net more "data-aware" is by realizing "per-flow" (rather than "per-packet") routing [15]. In ordinary per-packet routing, each and every incoming packet must be processed as described in Chap. 2. A processor in the router extracts the packet's destination address and queries the routing table to determine the next-hop address. Another processor (the switching fabric) puts the packet into a suitable output transmission queue. These two processors consume 80% of the power and space in the router.

On the other hand, with per-flow routing, we must only identify and process the first packet of any new data flow. The next-hop address is then stored in a hash table (coupled with the flow identifier), that is, a data structure that allows for a much faster lookup time than the routing table. Any subsequent packet of the flow is thus "switched" directly to the output buffer rather than having to be "routed." In fact, when this book was going to press, prototype per-flow routers had already shown to consume one-fifth of the power of a comparable per-packet router and to occupy a tenth of its space.

A further benefit of per-flow routing is that, thanks to its greater data awareness, it allows for more sensible congestion-control algorithms. For instance, the router can try to "detect" and "protect" the more sensitive "stream-based" flows.

New routing architectures such as the one behind per-flow routing imply huge deployment costs, as all existing routers (not just their software) would need to be replaced. Thus, network operators will embark on this kind of radical overhaul project only if they can see substantial economic benefits.

In the meantime, new paradigms keep emerging, providing further incentives in favor of innovation. Chapter 8 explores mechanisms through which we can publish data directly onto the network in order to pursue the dream of "data-aware" networks. Currently, the Net only understands about sources, destinations and distances. Routing tables are dumb; they cannot take into account the location and redundancy of data. By contrast, if the same file is stored in multiple servers and user's terminals, the Net should be able to choose the best source, with the goal to maximize network utilization. The challenge of data-aware networks is to route "information" rather than raw bits.

3.7 Problem 6: Inadequate Net-Search Engine

Thus far, we have established that the Net is a dumb, bit-crunching machine, incapable of making parallel transmissions. Is this really a problem? Is there any alarming sign suggesting that we need to imminently re-think the Net's protocols and its fundamental architecture?

There is an inexorable trend that should not escape our attention. Data sources are no longer confined within the boundaries of dedicated server machines. People have stopped being mere "consumers" of content; they "produce" and "broadcast" multimedia materials. PCs and cell phones "buffer" and "relay" data at an increasing pace. Thus, our information sources can reside virtually anywhere in the Net.

In today's digital society, the information has reached a "ubiquitous" status. Thanks to P2P IPTV frameworks, one can stream down directly from somebody else's PC, rather than connecting to the originating channel server (as discussed in Fig. 1.3). The same applies to music files, books and any other digital materials.

Currently, virtually anybody can "publish" information via the Net. However, the Net is oblivious to data. Before, we mentioned the benefits of making the Net more aware of the "data" dimension. The proliferation of P2P applications demonstrates that this transition is already underway. On the other hand, the proliferation and circulation of information cannot serve any purpose unless we can easily "discover" what is "relevant" to us.

Search engines play a pivotal role in this direction, but they can only capture the more "stable" data such as the files and streams that reside on servers. Conventional Internet search engines become ineffective when the information is "transient" and "volatile" (such as in P2P applications), or when it resides within the more "private" margins of our PCs and phones.

This new search dimension is explored in Chap. 9. We visit a futuristic scenario in which the data "plane" (the information) blends in with the network "plane." This is just another way of saying that the Net becomes data-aware. Our aim is to present mechanisms for Net-searching, i.e., methods for performing "deep" data discovery (beyond the servers and reaching into the user's terminal) via data-aware networks.

3.8 Concluding Remarks

The Net is far from being perfect; in fact, one can think of many ways in which it can be improved. Some observers think that the current incremental (evolutionary) approach—whereby technical limitations are addressed via "patches"—can still work for many years to come. By contrast, the "Future Internet" advocates believe that a complete architectural overhaul is urgently needed. The authors of this book have taken a more detached position: no matter how the Net will change (and it will undoubtedly change considerably), some new mechanisms will have to be realized. We have chosen six out of the many more possible ways in which the Net needs to be upgraded. These are not the only issues with the Net. Our choice was driven by two factors: urgency and availability of consolidated solutions.

References

1. Peterson LL, Davie BS (2007) Computer networks: a systems approach, 4th edn. Morgan Kaufmann, San Francisco
2. Miller FP, Vandome AF, McBrewster J (2010) Classless inter-domain routing. Alphascript Publishing
3. Sebesta RW (2010) Programming the World Wide Web, 6th edn. Addison Wesley, Boston
4. Marchese M (2007) QoS over heterogeneous networks. Wiley, Chichester
5. Soliman H (2004) Mobile IPv6: mobility in a wireless Internet. Addison Wesley Professional, Boston
6. Liotta A, Liotta A (2011) Privacy in pervasive systems: legal framework and regulatory challenges. Pervasive computing and communications design and deployment: technologies, trends and applications. Idea Group Publishing, London
7. Oram A (2001) Peer-to-peer harnessing the power of disruptive technologies. O'Reilly Media, Sebastopol
8. Handley M (2006) Why the Internet only just works. BT Technol J 24:119–129
9. Sarkar SK, Basavaraju TG, Puttamadappa C (2007) Ad hoc mobile wireless networks. CRC Press, Boca Raton
10. Alhaisoni M, Liotta A, Ghanbari M (2010) Resilient P2P streaming. Int J Adv Netw Serv 3:209–219
11. Agboma F, Liotta A (2010) The four facets of multimedia streaming. Next generation mobile networks and ubiquitous computing. IGI Global Publishing, Hershey, pp 59–68
12. Alhaisoni M, Liotta A, Ghanbari M (2010) Resource-awareness and trade-off optimization in P2P video streaming. Int J Adv Media Commun Special Iss High Qual Multimedia Stream P2P Environ 41:59–77
13. Barkai D (2002) Peer-to-peer computing: technologies for sharing and collaborating on the net. Intel Press, Hillsboro
14. Alhaisoni M, Ghanbari M, Liotta A (2010) Scalable P2P video streaming. Int J Bus Data Commun Netw 6:49–65
15. Roberts L (2009) A radical new router. Spectr IEEE 46:34–39

Chapter 4
Spontaneous Networks

Abstract How can we increase the capillarity of the Net without facing the daunting issues that come with large-scale infrastructures? Can we embed all necessary protocols into our terminals and then use the terminals themselves to relay packets? This chapter develops the vision of ubiquitous connectivity, pinpointing foundations and problems. Networks made without any dedicated hardware are possible, but require new protocols. Here, we discover how to build spontaneous, ad hoc networks starting from extremely simple mechanisms.

There are three kinds of death in this world. There's heart death, there's brain death, and there's being off the network
Dr. Guy Almes, Leader at Texas A&M University

4.1 The Gift of Ubiquity

At some point, ubiquitous connectivity will enable communications everywhere. Can this be achieved through a conventional infrastructure-based network? Even if we could prove that existing technologies would be able to work at such large scales, economical and business considerations suggest a negative answer. The investment required to deploy a capillary network is beyond imagination. Operational and maintenance costs would be astronomical.

A more promising approach is to use infrastructure-based networks to form the communication backbone, though adopting a different approach to increase the network capillarity. The basic idea is to extend the reach of existing networks by complementing them with so-called "infrastructure-less" networks. This does not mean that the infrastructure is eliminated; it is, in fact, embedded into the user's terminal.

A. Liotta and G. Exarchakos, *Networks for Pervasive Services*,
Lecture Notes in Electrical Engineering, 92, DOI: 10.1007/978-94-007-1473-1_4,

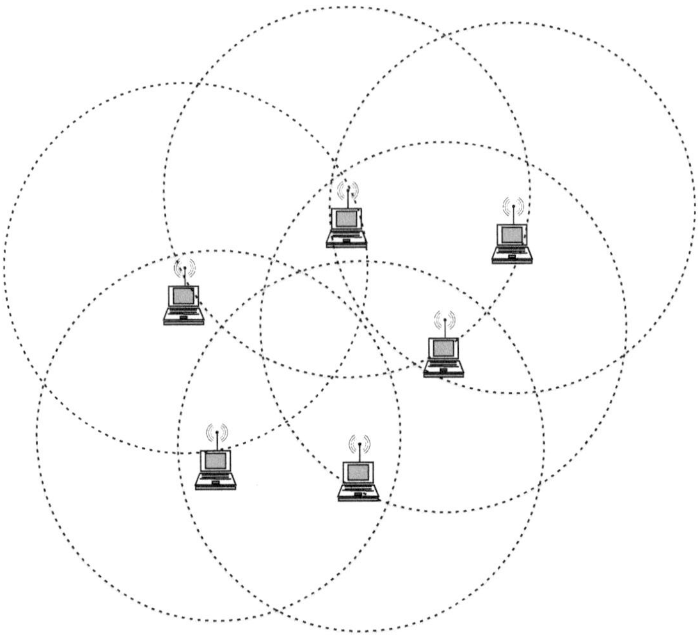

Fig. 4.1 An example of an autonomous network cluster

The Internet-enabled terminals of today are already powerful enough to carry out all three fundamental communications functions (*connect*, *path discovery* and *stay connected*). They can perform routing tasks just like ordinary routers, relaying packets on behalf of the other terminals. Thus, portable terminals can actually interconnect with each other and form autonomous network clusters (Fig. 4.1). In this way, a network can be formed "spontaneously" wherever the user goes. In addition, when at least one of the terminals has full Internet connectivity (via an infrastructure network), all terminals end up in the global Net (Fig. 4.2) [1].

Beyond extending the reach of infrastructure-based networks, ad hoc networks (this is just another term used to refer to *infrastructure-less* networks) [2] find numerous application domains [3]. They are invaluable in any situation that requires setting up a network "on the fly" or "on demand." For instance, an emergency-response team requires a reliable communication platform to better coordinate its intervention in a disaster area. A communication infrastructure may not exist or might have been damaged by the disaster. Thus, the ability to form a spontaneous network automatically has crucial importance.

Similarly, ad hoc networks are useful to create connectivity in remote areas. For instance, Residential Broadband Mesh Networks represent a much more economical alternative to conventional infrastructure networks. Hostile environments, such as military warzone areas, are yet another example.

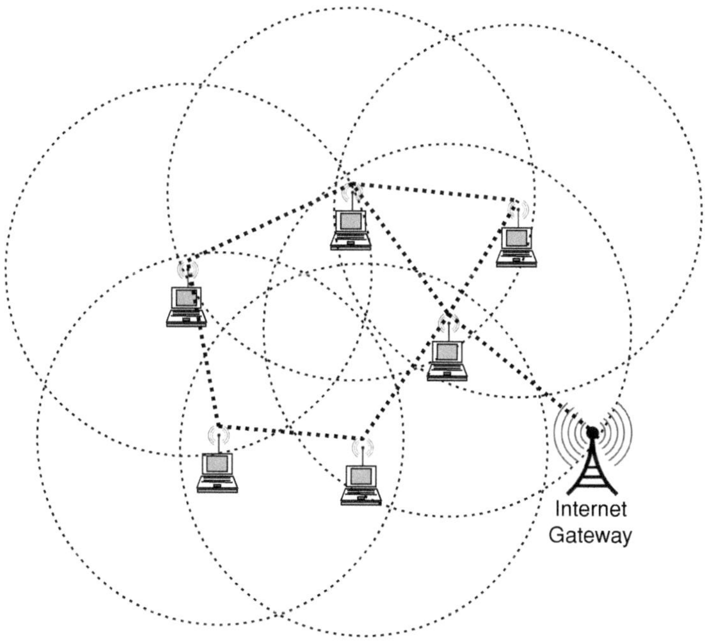

Fig. 4.2 The ad hoc network cluster gains internet connectivity via one of the terminals

Because of these attractive applications, ad hoc networks are gaining significant attention in the scientific community [4, 5]. However, and as it will become apparent in the course of this chapter, these networks have rather peculiar requirements which call for new routing protocols [2]. Before we look at the issues relating to the construction of routing paths, we must direct our attention to how neighboring ad hoc terminals "connect" to each other.

4.2 Spontaneous Connectivity

Before we can build the network, including routes and paths, we need to organize the communication within the individual links. In infrastructure networks, this aspect is transparent to the end user terminal. It is the routers who handle the contention problem which arises when more transmissions hit the same link.

By contrast, in an ad hoc network, each terminal must sense who his immediate neighbors are and learn how to "connect" to them. A simple bootstrapping protocol is depicted in Fig. 4.3.

We need to solve yet another problem. A, B and C are sharing the ether (the transmission channel); so if they use the same frequency and transmit at the same time, they might generate interference with detrimental consequences for the

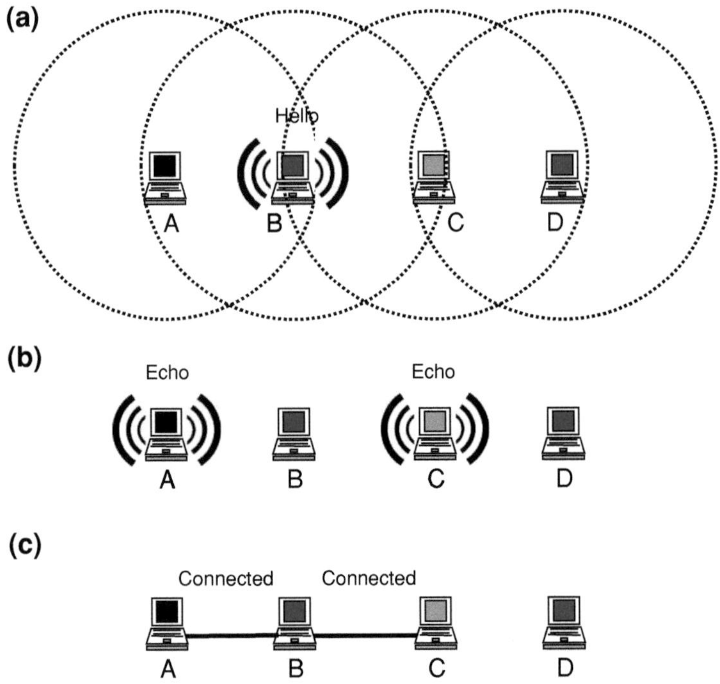

Fig. 4.3 Connectivity bootstrapping at terminal B: **a** terminal B broadcasts a *Hello* packet, **b** terminals A and C are within the transmission range of B, sense its advert and respond with an *Echo* packet, **c** B has now learned about A and C, and can connect to them

packets (packet collision problem) [6]. We now consider two typical collision problems and then introduce a suitable media access control protocol.

4.3 The Hidden-Terminal Problem

Suppose that A and C want to communicate with B (Fig. 4.4). A and C are out of transmission range so they have no way of coordinating their actions. If A transmits a frame (f_a) while C transmits a frame (f_c), B might not be able to receive anything (f_a and f_c collide). What is worse, neither A nor C have any way to detect the problem [3, 6, 7].

We could improve the protocol slightly by including acknowledgements. Any transmitter expects the recipient to send back an *Ack* signal as confirmation that the packet has been received. The transmitter will expect the *Ack* within a given time or, otherwise, assume that something has gone wrong with the original packet. In that case, it will just re-transmit the packet.

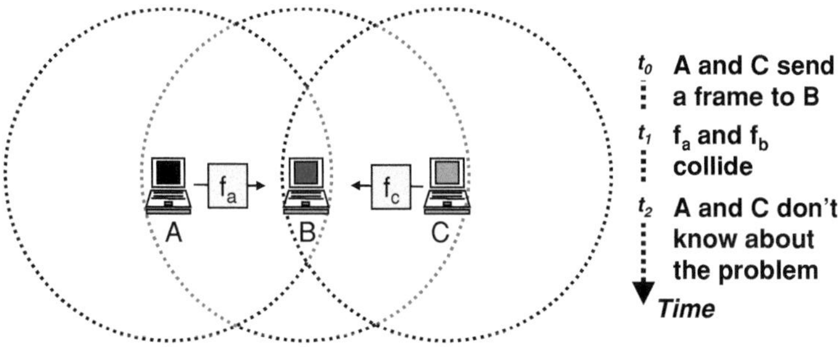

Fig. 4.4 The hidden-terminal problem

The *Ack* signals alone do not necessarily eliminate collision. Suppose A and C transmit simultaneously at time t_0 and, then, fail to receive their *Ack* within a time span Δ_t. At time $(t_0 + \Delta t)$, A and C retransmit, thus again generating collision. We can fix this glitch easily. Each transmitter waits for a further time ∂t which is determined by a random generator. By adding a stochastic delay, we reduce the probability that transmitters remain in sync.

Yet, how efficient is this transmission protocol? Whenever a collision takes place, we are adding a delay of $(t_0 + \Delta t + \partial t)$. As the packet transmission rate increases, so does the probability of collision. As the transmitter's density (the number of transmitters lying within the same transmission range) increases, so does the probability of collision. A significant portion of bandwidth goes wasted because we are not doing anything to prevent collision. Let us examine a second scenario before discovering the advantages of "collision avoidance" protocols.

4.4 The Exposed-Terminal Problem

Suppose B wants to communicate with A, and C with D (Fig. 4.5). Let us assume that B transmits first. Because C is within B's transmission range, it will be able to detect that B is transmitting (by sensing its carrier). Hence, in an attempt to prevent collision, C will defer its own transmission. However, this is the wrong decision: A and D are out of range so even if B and C were to transmit simultaneously, their frames f_b and f_c would not interfere at the receivers A and D, respectively. In this case, the fact that B and C are exposed to each other delays the transmission unnecessarily [3, 6, 7].

4.5 Preventive Measures to Avoid Collision

The media access protocols described so far in this chapter belong to the category of "random access" schemes—nodes are allowed to transmit pretty much at any time. These can offer relatively low throughput as they put minimal effort into

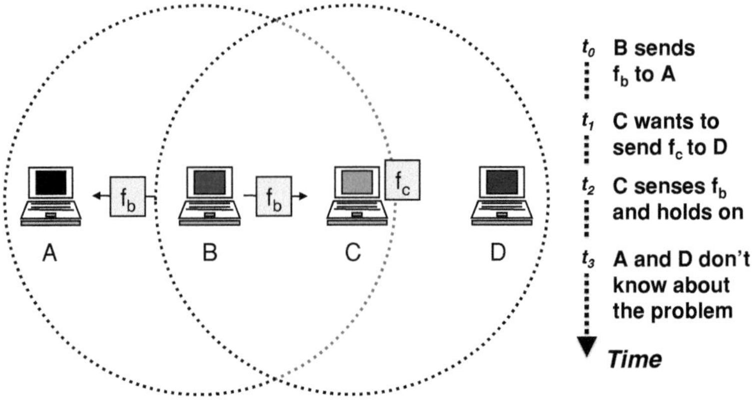

Fig. 4.5 The exposed-terminal problem

trying to prevent collision (they merely act after the collision has taken place). Also, we have seen with the examples of Figs. 4.4 and 4.5 that "carrier sensing" alone can achieve very little in a wireless environment since collisions can still occur at the receiving nodes.

A better performance is achieved with protocols that involve some form of dynamic channel reservation prior to transmission. Figure 4.6 illustrates how multiple access with collision avoidance (MACA) [3, 8] can address the hidden-terminal problem. The key idea is that terminals must go through an approval procedure before they are allowed to transmit data.

MACA employs two small signaling packets, the request-to-send (RTS) and the clear-to-send (CTS). Let us follow the sequence of events in Fig. 4.6. Terminals A and C intend to transmit data to B. Suppose that A takes the initiative first, sending the RTS. With the exception of the intended recipient B, any other node that overhears the RTS must refrain from transmitting packets for a sufficient time as to prevent the collision of signaling packets.

Having received the RTS, B replies by broadcasting a CTS. With the exception of the original transmitter A, any other node hearing a CTS must refrain from transmitting for the duration of A's transmission. At time t_2, node C hears the CTS and thereby remains silent until A finishes. Then, it initiates its own RTS/CTS procedure.

MACA can significantly improve the network throughput—the RTS/CTS sequence reduces the probability of data packets collisions. MACA, however, does not fully eliminate the hidden-terminal issue since it is still possible that the signaling packets collide. An example of such event is depicted in Fig. 4.7, whereby A and C happen to initiate the RTS/CTS sequence simultaneously.

It should be noted that the collision of signaling packets is far less problematic than that of data packets. Firstly, signaling packets are much smaller than data packets, so they are transmitted faster. In turn, the probability of collision is reduced. Furthermore, when signaling packets do collide, the impact on the

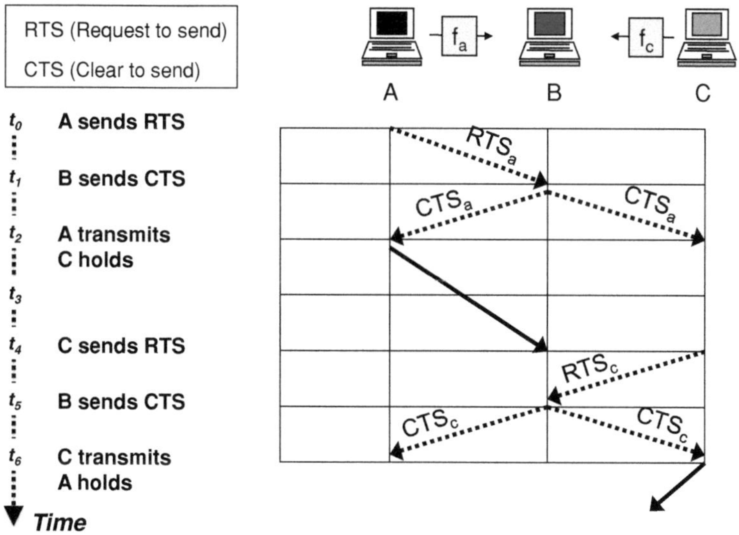

Fig. 4.6 Multiple access with collision avoidance (MACA) in a hidden-terminal scenario

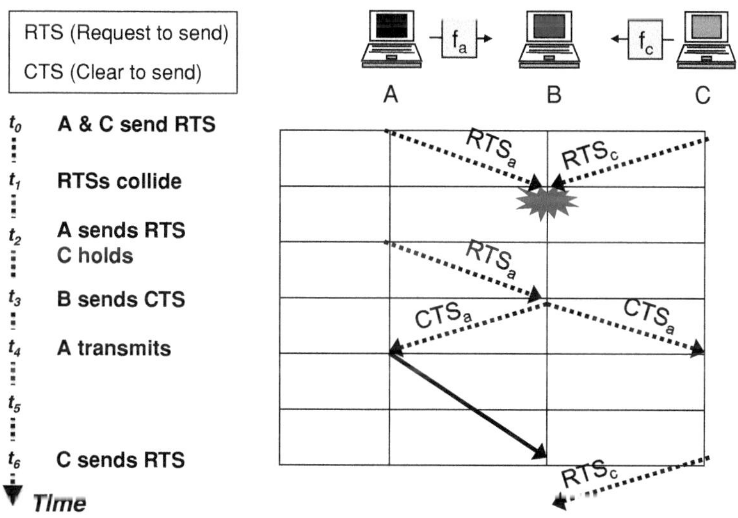

Fig. 4.7 MACA does not completely solve the hidden-terminal problem, though it improves it

network performance is negligible since the capacity wasted to re-transmit them is considerably smaller than the traffic incurred by data packets.

Let us now see how MACA addresses the exposed-terminal problem (Fig. 4.8). Terminals B and C want to transmit to A and D, respectively, which are out of each other's transmission range. Recall from Fig. 4.5 that C will sense B's carriers

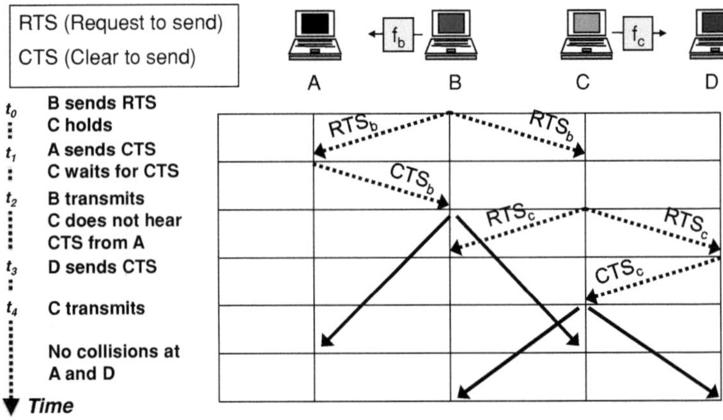

Fig. 4.8 MACA in an exposed-terminal scenario

and unnecessarily defer its transmission to prevent interference. By contrast, with MACA at time $t_{1,}$ C receives the B's RTS signal and continues to defer transmission, expecting to hear a CTS by time t_2. Yet, because C does not receive the CTS at t_2, it will correctly infer that the intended receiver (i.e., A) is out of range. Thus, C's transmission cannot interfere with the ongoing transmission by B. Consequently, C initiates its own channel reservation procedure at t_2, i.e., much earlier that it could do on a simpler "carrier sensing" scheme.

MACA is effective because signaling packets are significantly smaller and transit considerably faster than data packets. Hence, collisions tend to be less frequent and less expensive. However, scientists are still dedicating great attention to the many ways in which MACA could be further improved, particularly in the context of ad hoc networks. We conclude with another scenario illustrating a data packet collision (Fig. 4.9).

4.6 Path Discovery in a Volatile Networks

We have now explored how the first of the three fundamental communication principles applies to ad hoc networks. Let us assume that terminals are able to find their neighbors and "connect" to them. We still need to "discover" end-to-end paths and "stay connected" while those terminals move freely.

Can we just adopt ordinary routing protocols such as those of Chap. 2? We did state before that Link State is a general-purpose protocol; but ad hoc networks are incredibly more volatile than typical IP networks. In the Net, the routing function is performed by dedicated hardware (the routers) that is not moving. Routing paths do have to be computed dynamically, but this is needed mainly to cater for the changeable nature of traffic in intensity, origination and destination. In the Net,

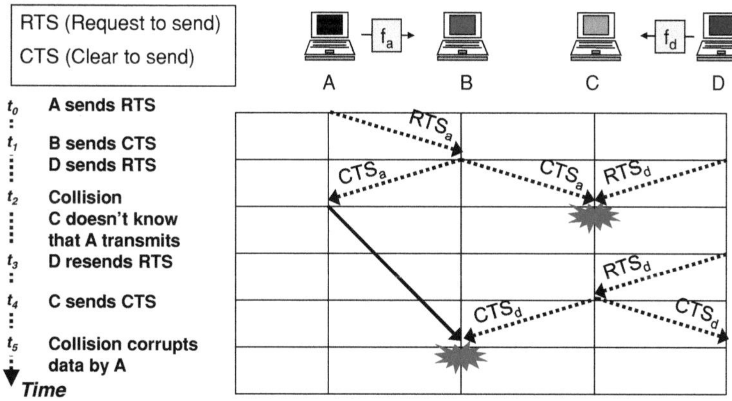

Fig. 4.9 MACA reduces the probability and the cost of collisions but cannot always eliminate the problem

the data transport topology varies for the purpose of load balancing. However, the physical topology does not change much.

In contrast, in ad hoc networks the routing network is made of "mobile" nodes, the user's terminals. Hence, in addition to the transport topology, the physical topology continuously changes as people move around. Figure 4.10 illustrates how even the slightest terminal movements may result in a radically new topology.

Thanks to a broad range of scientific studies, the reasons why conventional routing schemes do not perform well on ad hoc networks are well known. In short, these are

- Higher dynamics (physical-as well as transport-level dynamics).
- Higher packet loss (media access protocols incur high error rates).
- Routing nodes have considerable constraints in terms of energy (battery-operated) and computational power and memory).
- The increased signaling rates that would be required to stabilize the network cannot be sustained by network (traffic overheads) and terminals (processing overheads).

Let us now initiate the journey towards ad hoc routing by starting off with a simple approach.

4.7 The KISS Approach

As network dynamics increase, the task of maintaining efficient routes becomes overly intrusive. At some point, the amount of signaling required to keep up with topological changes overwhelms the network. When this occurs, *data broadcasting* may represent a competitive alternative to the more conventional routing schemes. Instead of spending network and computational resources to build up the routing

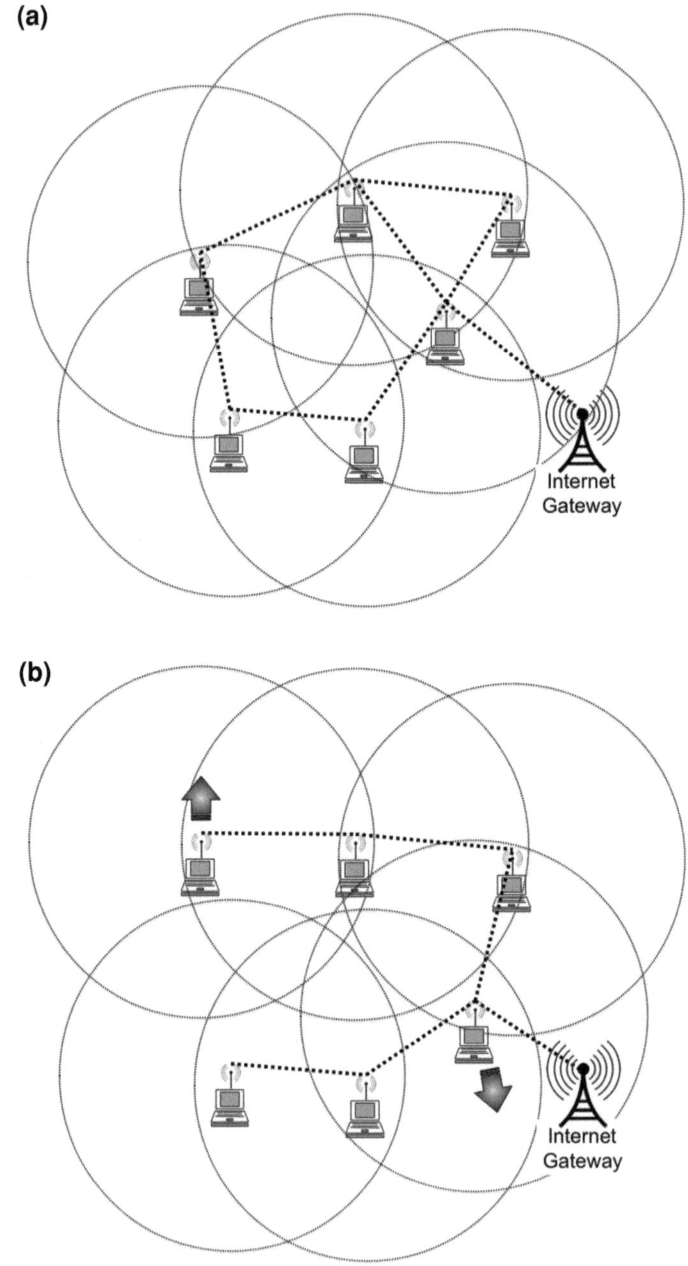

Fig. 4.10 **a** Network topology before node movement, **b** two terminals move slightly, causing a radical change of the topology

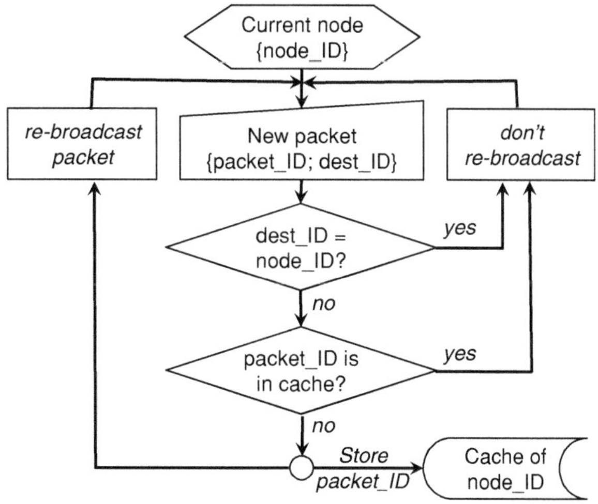

Fig. 4.11 Behavior of each of the nodes in a broadcast-based transport network

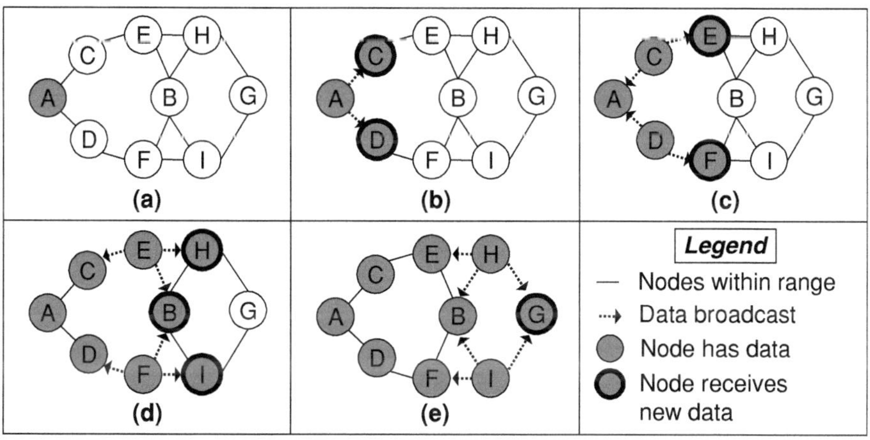

Fig. 4.12 Broadcast-based transport from nodes A to B in a 9-node topology: **a** initial status, **b** node A broadcasts the packet, but only C and D are within transmission range, **c** nodes C and D re-broadcast the packet as soon as they receive it. Two packets are sent back to A unnecessarily. **d** Nodes E and F re-broadcast the packet. Four packets are sent unnecessarily, respectively, to C, D, H and I. **e** Node B is the intended destination so it does not re-broadcast the packet. Nodes H and I rebroadcast six packets unnecessarily

tables, we can directly broadcast the packets, one at time, from the source. Thus, when a node receives a new packet, it will simply react as depicted in Fig. 4.11, very much in line with the KISS (keep it simple & stupid) design principle.

Figure 4.12 depicts the overall process, assuming that node A is the source and node B is the destination. The network nodes do not have any routing table, though they can keep track of the packets that transit through them (each packet has a unique ID) in order to contain the broadcasting process. This approach comes with many benefits:

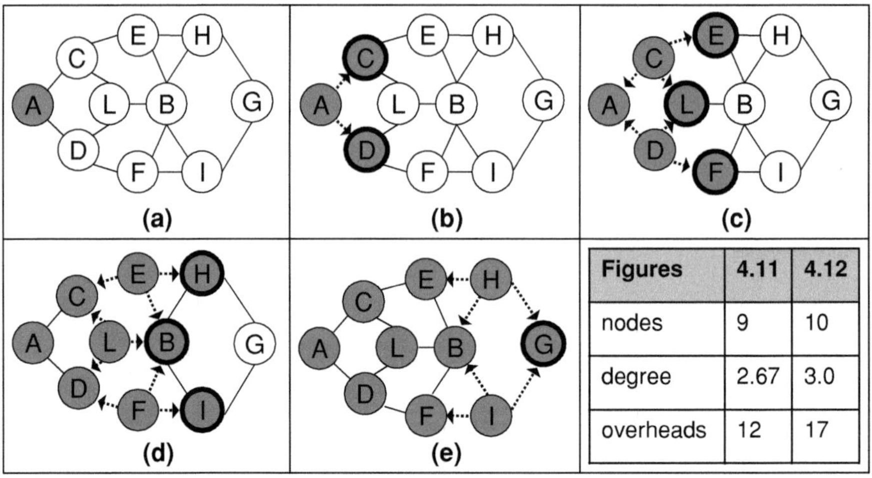

Figures	4.11	4.12
nodes	9	10
degree	2.67	3.0
overheads	12	17

Fig. 4.13 Broadcast-based transport from nodes A to B in a 10-node topology. In comparison with Fig. 4.12 this topology only has one extra node, but generates 17 unnecessary packets (instead of 12)

- No signaling overheads.
- Utmost simplicity.
- Minimal processing (essential in battery-operated terminals).
- High reliability (thanks to redundant transmission).

On the other hand, broadcast-based transport incurs substantial data-transmission overheads since each existing path between source and destination delivers duplicate packets. In addition, the transmission process does not usually stop when the packet reaches its intended destination. This can be observed, for instance, in Fig. 4.12: step (e) is unnecessary since the packet has already reached destination B.

As the number or nodes increases, so do the overheads. By comparing the two topologies of Figs. 4.12 and 4.13, we can see that the addition of just a single node brings the number of overhead packets from 12 up to 17.

With broadcast-based transport, too many nodes receive and relay packets. Thus, when the scale and density of the system increases, we must return to the routing strategies. Before any data packet is injected into the network, control packets are used to discover the routes. The following chapters will offer plenty of opportunities to find out how to overcome the limitations of ordinary routing protocols.

References

1. Hekmat R (2006) AD-Hoc networks: fundamental properties and network topologies. Springer
2. Qadri N, Liotta A (2009) Analysis of pervasive mobile Ad Hoc routing protocols. Pervasive computing. Innovations in intelligent multimedia and applications. Springer, London, pp 433–453

3. Sarkar SK, Basavaraju TG, Puttamadappa C (2007) Ad hoc mobile wireless networks. CRC
4. Qadri N, Liotta A (2009) Effective video streaming using Mesh P2P with MDC over MANETS. J Mobile Multimedia 5:301–316
5. Qadri N, Liotta A (2010) Peer-to-peer over mobile ad hoc networks. Next generation mobile networks and ubiquitous computing. IGI Global Publishing, pp 105–121
6. Mueller A (2007) Efficient wireless MAC protocols-analyzing quality of service in wireless networks. VDM Verlag Dr Mueller eK
7. Tanenbaum AS (2002) Computer networks. Prentice Hall
8. Wang G (2008) MAC layer and routing protocols for heterogeneous ad hoc networks. VDM Verlag Dr Muller Aktiengesellschaft & Co KG

Chapter 5
Reactive Networks

Abstract Networks strive to keep "all" of their nodes connected. However, is this really necessary? Do we actually need to "continuously" maintain routes from and to "any" possible destination? This chapter looks at networks that can discern between active and non-active paths. The idea is to care for the nodes that are actively intercommunicating, leaving the rest of the network in standby mode. In this chapter, we will explore on-demand routing, one of the key ingredients that can make networks more reactive on a larger scale.

Doing nothing is better than being busy doing nothing
Lao Tzu, Philosopher (founder of Taoism)

5.1 Why Networks on Demand?

The Net is a massive system that spends enormous amounts of energy on trying to remain connected. The routing protocols revised in Chap. 2, distance vector (DV) and Link State (LS), strive to maintain routes from any point to any other point. Any path is equal, regardless of whether or not it is in use. Routes exist and are re-calculated even when they lead to idle terminals. The same amount of signaling is dedicated to every node, not considering whether they are static or mobile.

To better appreciate why this "brute-force" approach leads to a slow-adapting Net, consider the simple network of Fig. 5.1. Recall from Chap. 2 that every router will have its own routing table; each table will have a number of entries equal to the number of visible terminals (or networks); and that to achieve a fully operational network, all those tables must be kept up to date. Thus, the simple nine-node topology of Fig. 5.1 will generate nine tables, containing nine entries each. A total

A. Liotta and G. Exarchakos, *Networks for Pervasive Services*,
Lecture Notes in Electrical Engineering, 92, DOI: 10.1007/978-94-007-1473-1_5,

of $9 \times 9 = 81$ entries must be continuously maintained even when nobody is transmitting any data.

Real networks are far bigger than our sample nine-node topology. Furthermore, if every terminal can perform routing functions (as in ad hoc networks), the number of table entries will further increase. Add to this the possibility that terminals can move freely and the picture is complete: routes become obsolete even before the network can update them.

In the sections that follow, we shall see that networks can function even in the absence of any pre-computed paths [1]. By keeping the routing tables to a bare minimum and introducing mechanisms in order to build paths "on demand," the resulting network will become far more "reactive." The key idea is to identify the paths that are more active and then prioritize their maintenance.

5.2 A Traffic-Free Network

If nobody is injecting any traffic in the network, we do not need to build any paths. However, we still need to keep the network ready for any "path discovery" request [2]. The very first task is to "bootstrap" the network, making sure that every node is aware of its actual neighbors. Every node must continuously perform a "neighbor discovery" process in order to maintain connectivity with its immediate neighbors. Until a node has discovered the neighbors, its routing table remains empty. In Fig. 5.2, we can follow the neighbor discovery process of node A. Eventually, the table of node A is populated with two entries (one per neighbor), while nodes B and D have only one entry (neighbor) each.

If nobody transmits, the network can remain in "standby" mode. The only activity is "neighbor discovery," which merely aims to keep nodes connected to their immediate neighbors. The routing tables are now only populated with neighbor-node entries; so they are much smaller than the tables of a conventional routing protocol and do not grow with the number of nodes. For instance, in our "standby" network, we only need to maintain 24 table entries (Fig. 5.3), while a comparable DV routing network would operate on 81 (Fig. 5.1).

Thus, neighbor discovery is merely a background process that incurs minimal signaling overheads. However, we still need to build complete paths; though not until some node decides to transmit data.

5.3 Our First Path

We now want to inject traffic into the traffic-free network of Fig. 5.3 Assume we want to transmit between nodes A and B. Node A does not have any entry for destination B, so it triggers the "direct path" discovery process depicted in Fig. 5.4 Node A broadcasts a Route Request (RREQ) packet, indicating B as the intended destination. While the RREQ starts flooding the network, the nodes that receive it

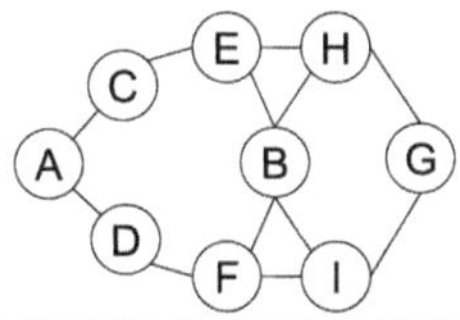

Routing table of node A

Dest.	Next-hop	Distance
A	-	0
B	C	3
C	C	1
D	D	1
E	C	2
F	D	2
G	D	4
H	C	3
I	D	3

Routing table of node B

Dest.	Next-hop	Distance
A	E	3
B	-	0
C	E	2
D	F	2
E	E	1
F	F	1
G	H	2
H	H	1
I	I	1

Routing table of node C

Dest.	Next-hop	Distance
A	A	1
B	E	2
C	-	0
D	A	2
E	E	1
F	A	3
G	E	3
H	E	2
I	E	3

Routing table of node D

Dest.	Next-hop	Distance
A	A	1
B	F	2
C	A	2
D	-	0
E	F	3
F	F	1
G	F	3
H	F	3
I	F	2

Routing table of node E

Dest.	Next-hop	Distance
A	C	2
B	B	1
C	C	1
D	C	3
E	-	0
F	B	2
G	H	2
H	H	1
I	B	2

Routing table of node F

Dest.	Next-hop	Distance
A	D	2
B	B	1
C	D	3
D	D	1
E	B	2
F	-	0
G	I	2
H	B	2
I	I	1

Routing table of node G

Dest.	Next-hop	Distance
A	H	4
B	H	2
C	H	3
D	I	3
F	H	2
F	I	2
G	-	0
H	H	1
I	I	I

Routing table of node H

Dest.	Next-hop	Distance
A	E	3
B	B	1
C	E	2
D	B	3
E	E	1
F	B	2
C	G	1
H	-	0
I	G	?

Routing table of node I

Dest.	Next-hop	Distance
A	F	3
B	B	1
C	B	3
D	F	2
E	B	2
F	F	1
G	G	1
H	G	2
I	-	0

Fig. 5.1 Routing tables in a conventional DV network

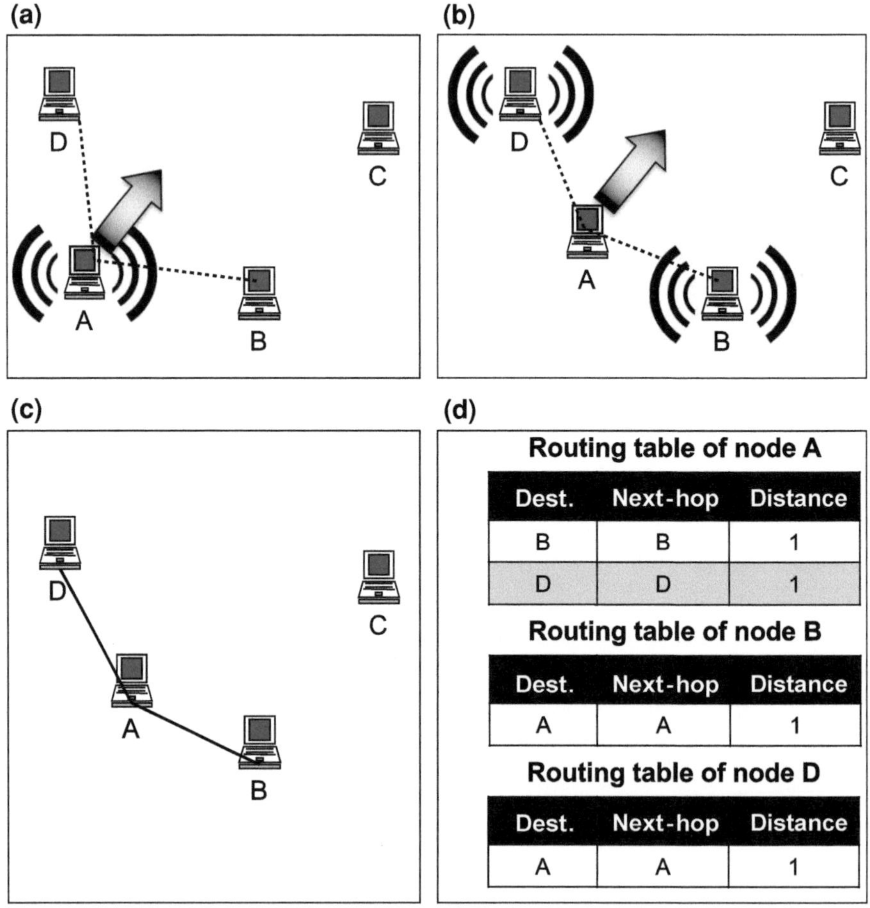

Fig. 5.2 Network bootstrapping from the point of view of node A; **a** while it is broadcasting a *Hello* packet, node A enters the transmission range of B and D, **b** as soon as they sense the *Hello* packet, nodes B and D respond with *Echo* packets, **c** node A can now add two new entries to its table, **d** after nodes B and D go through a similar hello-response process, their tables are populated with a single entry

start building a reverse path, leading back to the transmitter, A. From the node's point of view, a reverse path is structurally identical to a direct path. Each link of the path corresponds to a specific table entry. For instance, the reverse path D → A depicted in Fig. 5.4c corresponds to entry {A; A; 1} in the routing table of node D.

The RREQ flooding process continues until node B receives the RREQ (step (d)). At this point, node B sends a Route Reply (RREP) packet back to A, following the existing reverse path (step (e)). As the RREP visits the nodes between B and A, the direct path is gradually built (steps (f) and (g)).

In the meantime, the original RREQ continue to propagate via H and I until all nodes have been visited (step (e)). This protocol is not sufficiently sophisticated

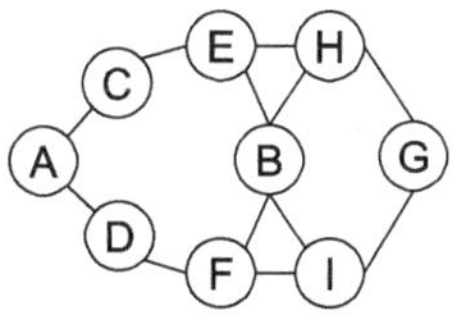

Routing table of node A

Dest.	Next-hop	Distance
C	C	1
D	D	1

Routing table of node B

Dest.	Next-hop	Distance
E	E	1
F	F	1
H	H	1
I	I	1

Routing table of node C

Dest.	Next-hop	Distance
A	A	1
E	E	1

Routing table of node D

Dest.	Next-hop	Distance
A	A	1
F	F	1

Routing table of node E

Dest.	Next-hop	Distance
B	B	1
C	C	1
H	H	1

Routing table of node F

Dest.	Next-hop	Distance
B	B	1
D	D	1
I	I	1

Routing table of node G

Dest.	Next-hop	Distance
H	H	1
I	I	1

Routing table of node H

Dest.	Next-hop	Distance
D	B	1
E	E	1
G	G	1

Routing table of node I

Dest.	Next-hop	Distance
B	B	1
F	F	1
G	G	1

Fig. 5.3 In a traffic-free network, the nodes keep track of only their immediate neighbors

enough to interrupt the RREQ broadcasting process as soon as the intended destination has been discovered. However, nodes can recognize duplicate or looping RREQs.

Figure 5.4h shows two valid paths. Note that reverse paths are built only to support the direct path construction and expire soon after step (g). If node B intends to communicate back to A, it will be necessary to initiate a new path discovery process. The final routing tables of nodes A and B are shown in (i) and (j), respectively.

5.4 Path Management

When paths are built on demand, we also need mechanisms to keep them up to date. It is not uncommon that an existing path becomes suboptimal or invalid due to node mobility. Moreover, better paths may appear at any time. We also need to implement a "garbage collection" scheme to ensure that all unutilized paths are pruned timely. Because of that, paths are created with an expiration time and the routing tables must contain some new fields:

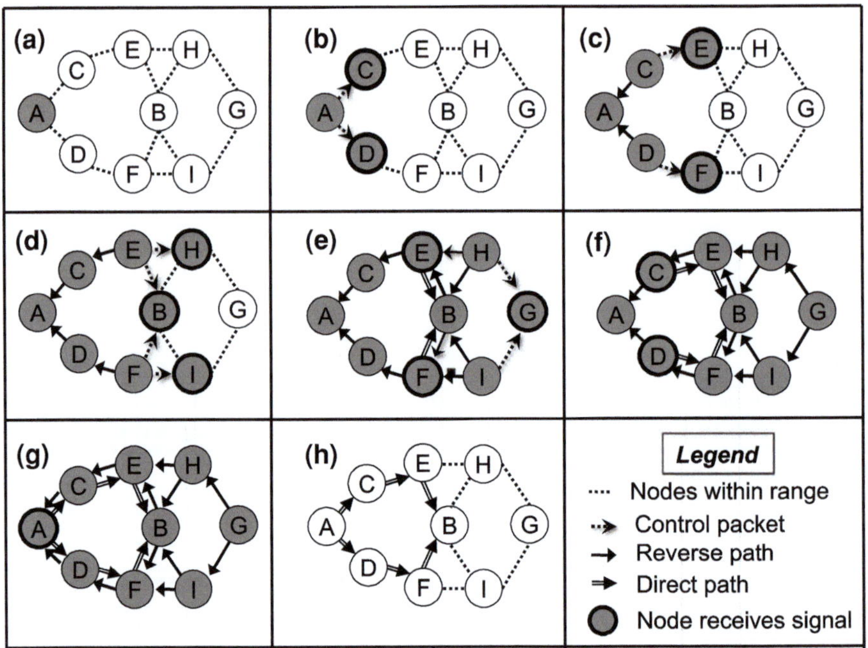

(i) **Routing table of node A**

Dest.	Next-hop	Distance
B	C	3
C	C	1
D	D	1

(j) **Routing table of node B**

Dest.	Next-hop	Distance
E	E	1
F	F	1
H	H	1
I	I	1

Fig. 5.4 On-demand path construction

- Sequence counter of the destination node: used to handle out-of-order signals, ensuring the creation of loop-free routes.
- List of active neighbors involved in the route table entry: used to rapidly tear down a path that has become invalid for some reason.
- Expiration time of the route table entry: this is reset up to *maximum_timeout* whenever the route entry is used to transmit data.

Likewise, any other routing protocol, the routing tables are orchestrated by the signaling packets. It is easier to understand the table management process if we follow a simple path discovery example, as shown in Fig. 5.5. Step (a) depicts the initial status. Node A has already discovered neighbors B and D, but now it wants to communicate with C. Path discovery starts with the creation of an RREQ packet that carries the following information:

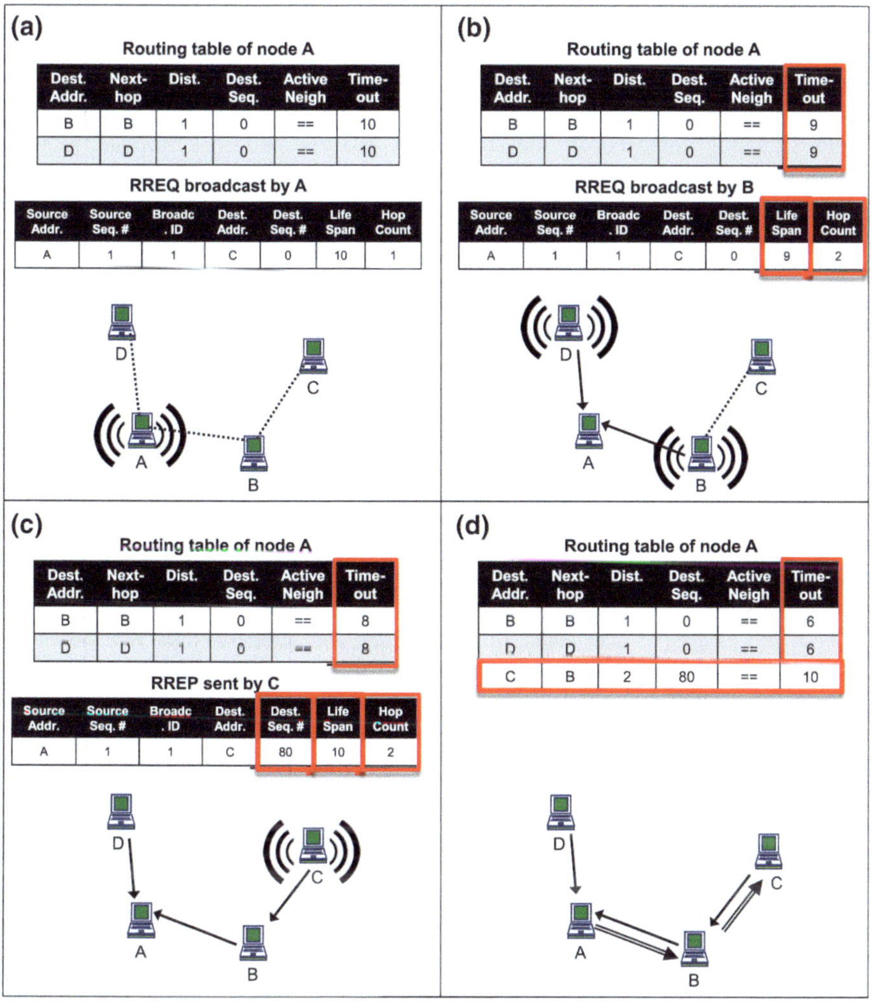

Fig. 5.5 Signaling packets RREQ and RREP are used to manage the routing tables

- *Source address* The initiator of the path discovery process.
- *Source sequence counter* Establishes the freshness of the information coming from the source. Its value is incremented when the source relays any signaling packet.
- *Broadcast identifier* A counter incremented each time the source issues a new RREQ.
- *Destination address* The end point of the path to be discovered.
- *Destination sequence counter* Used to maintain freshness information about the path (a higher sequence counter identifies a fresher route). Its value is incremented whenever the destination relays a signaling packet.

- *Lifespan* A timer (or counter) determining the length at which the signaling packet is kept alive before being purged from the network. As soon as the packet reaches a node, the *lifespan* value is decremented. If the resulting value is zero, the packet is pruned. This simple stratagem is used to contain the signaling incurred by the broadcast process and to eliminate looping packets. In fact, the *Hello* packets used in "neighbor discovery" are set with a *lifespan* equal to one because they are meant to only reach their immediate neighbors.
- *Hop count* Keeps track of the number of nodes visited by the signaling packet before reaching the destination. This value is used to set the *distance* field of the routing tables.

Figure 5.5b shows the snapshot when nodes B and D receive the RREQ from A. Assume that neither B nor D have a route entry for destination C. They set up a reverse path (pointing to A) and reprocess the RREQ. They decrement the *lifespan* counter and increment the *hop count*. Since the *lifespan* is still greater than zero, the RREQ is allowed to be rebroadcast. No new routes have been discovered at this stage, so the routing table of node A remains almost unchanged. The *time-out* values of each table entry are decremented, but are still above zero; thus, no entries need to be pruned.

To avoid excessive overheads, this *rebroadcasting* process stops as soon as the *lifespan* reaches a zero value. In fact, if we set the initial value of the *lifespan* too small in comparison to the network size, there is the risk that the path discovery process is interrupted before the destination can be reached by the RREQ. By contrast, if *lifespan* is too high, we end up injecting too many signaling packets. A possible solution is to start the discovery process with a relatively low *lifespan*. If the initiator does not receive any reply within a period of about twice the *lifespan*, it will restart the process using a higher *lifespan* value.

In Fig. 5.5c we capture the moment in which the intended destination, node C, receives the RREQ. Node C builds a reverse path to B and creates an RREP packet for A. The RREP contains the current *destination sequence counter* of node C and the *hop count* value received by the RREQ. As for any new signaling packet, the *lifespan* counter is reset. The RREP is then sent back to A via the reverse path. The routing table of node A still remains unchanged, that is, except for the *time-out* values that continue to decrease.

In our scenario, the RREP visits node B before reaching node A. At node B, the RREP triggers the generation of a new route entry {C; C; 1; 80; ==; 10} (not shown in Fig. 5.5). In the meantime, the routing table of node A further decrements their *time-out* values (down to 7). Then, the RREP is sent to A after its *lifespan* is also decremented (down to nine).

Finally, node A receives the RREP, acquiring all of the information needed to build a route entry for C. In Fig. 5.5d, we see the network status after the whole discovery process is complete. The reverse path is only needed to carry RREP packets, so it is set to expire more quickly than the direct path. The latter will be pruned after 10 s unless data packets transit through it.

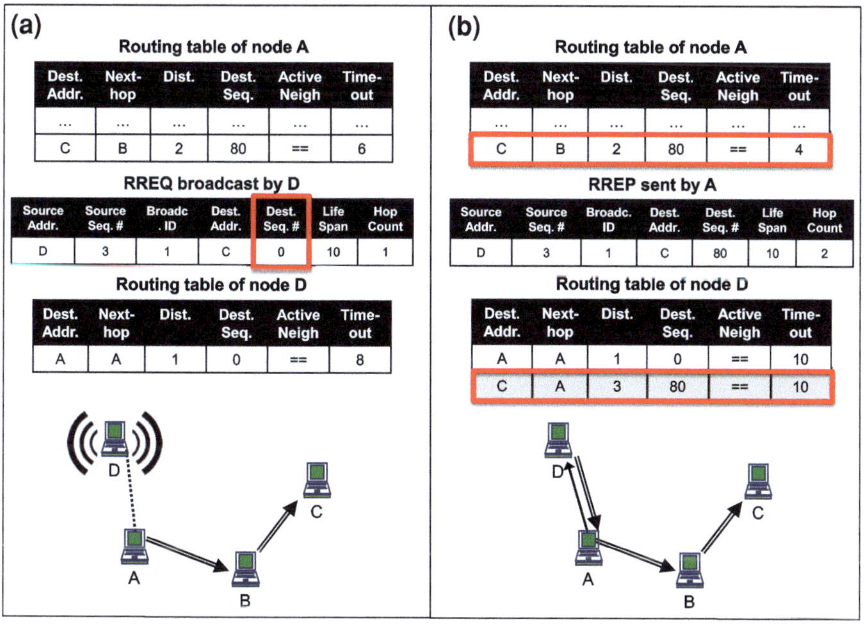

Fig. 5.6 Path discovery is quicker when part of the path already exists

5.5 Our Second Path

Building all paths from scratch would not be efficient. In fact, RREQ and RREP signals can do much more than just building paths on demand: they can also "expand" paths on demand, as exemplified in Fig. 5.6. Assume that the path A → B → C is still active when node D decides to transmit to C (step (a)). At this point, D has no clues as to how to reach C; therefore, it initiates a new path discovery procedure. The *destination sequence counter* of the RREQ is set to zero to signify that D has never received any information about this destination.

Upon receiving the RREQ, node A sets a reverse path (A → D) and realizes that its routing table already has an entry for destination C (step (b)). Consequently, node A interrupts the RREQ broadcast and sends an RREP back to A, including the path information. Node D can then create a new routing entry to destination C.

5.6 Global Synchronization

Thus far, we have analyzed scenarios only involving one transmitter at a time. When more nodes concurrently initiate transmission, different RREQs and RREPs might bring inconsistent information onto the same node. To resolve conflicts, we need some form of global synchronization to establish the freshness of the status

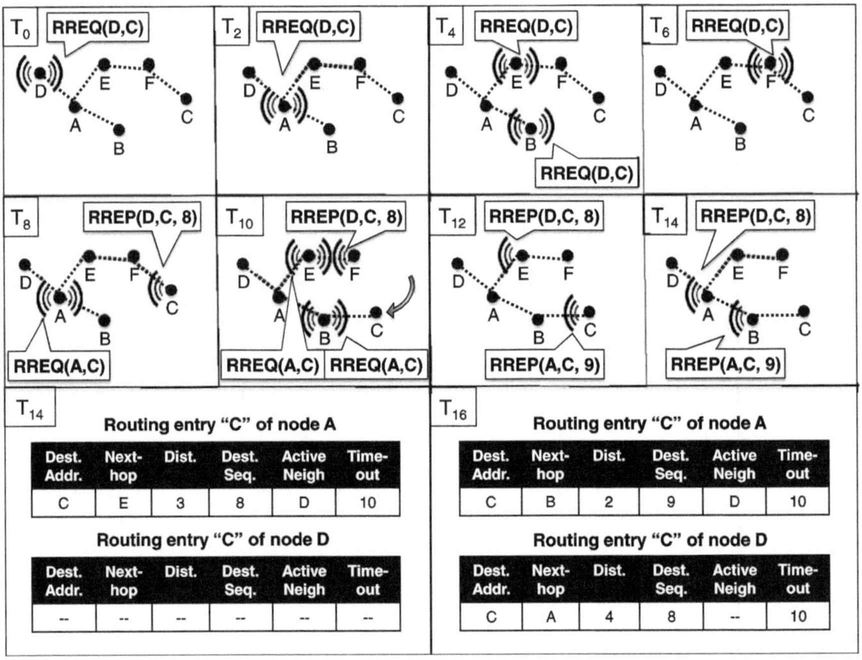

Fig. 5.7 Dynamic path updating

information carried by the signalling packets. To achieve this purpose, each node maintains two counters, the *node sequence counter* and the *broadcast ID number*. Both of these are used to time-stamp the signalling packets.

Let us observe the mechanics of time-stamping through the example of Fig. 5.7. We have two transmitters, D and A, which initiate path discovery at time T_0 and T_8, respectively. For simplicity, let us assume that the signaling packets take 2 s to travel between two adjacent nodes. The RREQ transmitted by D (RREQ(D, C)) reaches its destination, C at time T_8, while A is broadcasting its own RREQ(A, C). Node C now creates RREP(D, C, 8)—whereby the third parameter is the sequence counter of node C—and sends it back to D.

To make things more realistic, suppose that node C is moving and that at time $T_{10,}$ it hands over from C to B. Upon this topological change, the RREP(D, C, 8) becomes obsolete. Yet, there is no way to prevent this outdated information from propagating through the D $\rightarrow$ C reverse path (i.e., F, E, A, D).

Nevertheless, the network is actually capable of detecting this inconsistency and building the correct path. At time T_{14}, RREP(D, C, 8) reaches node A, generating an obsolete entry for destination C. However, the fresher RREP(A, C, 9) is already on its way to put things right: at time T_{16}, it reaches node A, generating an up-to-date entry for destination C.

This simple example shows that we can keep nodes in synch even in the absence of a global clock. Through time-stamps, we keep track of the relative time between subsequent signals. For instance, RREP(A, C, 9) is fresher than RREP (D, C, 8); therefore, the path A → B → C will override path A → E → C.

5.7 Error Management

The task of maintaining a globally consistent network is hindered by the transport latency of the signalling packets. The more dynamic the network, the higher the probability that RREQs and RREPs carry outdated information, leading to incorrect paths. Incorrect paths represent a real problem since they lead to performance degradation. However, engineering an "error-free" network is excessively difficult due to the many independent variables that influence its operation. Any node can start transmitting, receiving and relaying packets at any time. It may move unpredictably, causing dramatic topological changes, loops and dead ends.

A more effective alternative is to incorporate an error management system, capable of "catching" and "handling" errors. Simple error messages can help to address a number of problems, as exemplified in Fig. 5.8. Suppose we are still observing the scenario depicted in Fig. 5.7, intercepting it at time T_{13}. Node F issues one of its periodic *Hello* messages and then expects an *Echo* response within four time slots (twice the time for a signaling packet to travel between two adjacent nodes). Obviously, there will not be any *Echo* packet because node C is no longer within the transmission range of F. Thus, at time T_{17}, node F becomes aware that the table entry with destination C is incorrect and creates a route error message (RERR(C—A, D, E)) to inform all the active nodes (A, D and E), i.e., those nodes who relied on F to reach C.

To make things more entangled, suppose that node E starts transmitting data to C at time T_{18}, while the RERR is still traveling in the link F → E. Considering its routing table, node E relays the data via node F, unaware that F is now a dead end. At time T_{19}, our RERR reaches E and deletes the erroneous route to C (Fig. 5.8b). Unfortunately this is too late to catch the data packets that have already left node E; but the RERR puts things right for any data transiting via E after time T_{19}. In fact, the RERR fixes tables of A and D at times T_{21} and T_{23}, respectively.

This error message belongs to the category of "garbage collection" mechanisms, whose task is to delete paths that have become obsolete. Thanks to the simple scenario of Fig. 5.8, we can now appreciate that "garbage collection" addresses only part of the problem. Our DATA(E, C) is still travelling unnecessarily and, what is worse, node E is not aware of the problem.

We need a second kind of error message that can "catch" stray data, "inform" the originator and "rebuild" the outdated paths. To understand how to achieve this, let us follow the sequence of events from time T_{21}. As soon as DATA(E, C) reaches F, the node realizes that the sender (E) must have an incorrect path. Node F generates an RERR(C–E) which instructs E to initiate a new path discovery process. In fact, E issues a RREQ(E, C) as soon as it receives the RERR (at time T_{23}).

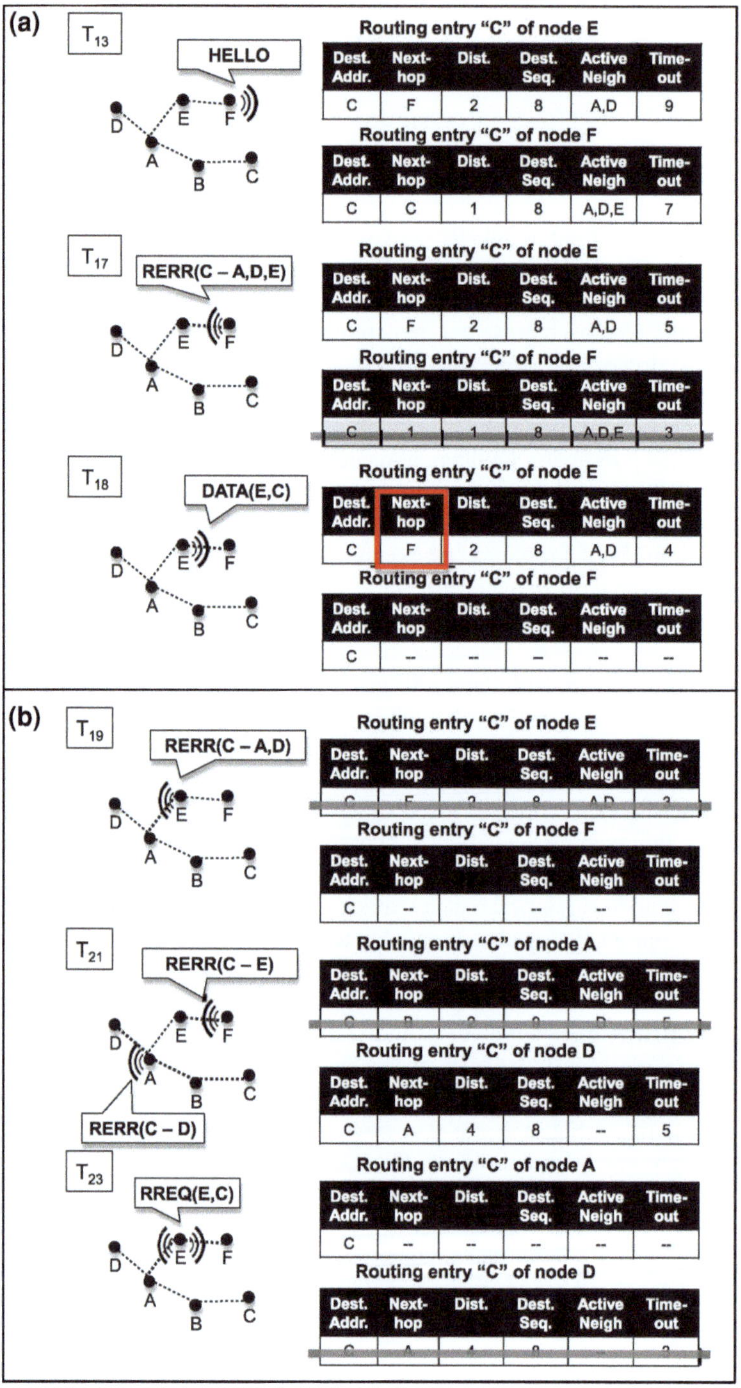

Fig. 5.8 **a** Error messages to manage erroneous paths, **b** error messages to manage erroneous paths

The two route error messages considered in this example work in perfect harmony. By time T_{23}, the first RERR manages to completely clear the network of all outdated paths to C. On the other hand, the second RERR takes care of re-building the path E $\rightarrow$ C to cater for the active transmitter E.

5.8 Remarks on Reactive Networks

A "reactive" network can promptly adjust to any external force, be it an application, the mobility of a node, or the migration of a cluster of nodes [3–6]. "Reactivity" represents the foundation of modern networks which are required to operate under an incredible variety of stimuli. To underline the importance of reactive networks, this chapter has explored the extreme case of a network that can function even in the absence of any pre-computed paths. This is a network that can operate only if it is able to react to stimuli. Such a network finds its roots in the context of the Ad hoc on-demand distance vector (AODV) routing protocol which was proposed by Perkings et al. in 2003 [7–11], which we have adopted in this chapter.

Networks that are merely reactive such as the AODV network teach us how to minimize overheads. When the network is in standby mode (no data traffic), the tables are virtually empty and the signaling messages only perform neighbor discovery.

However, the extreme reduction of overheads does not come without surprises. Since its conception, AODV has been the subject of intensive studies which have unveiled the limitations of a "purely reactive" approach [12]. Ironically, if the paths can only be built on demand, the network acquires an intrinsic latency: it is not able to respond before the completion of "path discovery" (obviously, this latency increases with the network diameter). Also, when the routes are used intermittently, the paths may expire in-between intermittent transmissions and will have to be re-built unnecessarily.

Therefore, on-demand networking should be regarded as just one of the ingredients of a reactive network. Another important element is pro-activity which is the subject of the next chapter.

References

1. Hekmat R (2006) Ad-hoc networks: fundamental properties and network topologies. Springer
2. Wang G (2008) Mac layer and routing protocols for heterogeneous Ad-hoc Networks. VDM Verlag Dr Muller Aktiengesellschaft Co KG
3. Sarkar SK, Basavaraju TG, Puttamadappa C (2007) Ad hoc mobile wireless networks. CRC
4. Mian A, Baldoni R, Beraldi R (2009) A survey of service discovery protocols in multihop mobile Ad-hoc networks. Pervasive Comput IEEE 8:66–74
5. Tonguz O, Ferrari G (2006) Ad Hoc wireless networks: a communication-theoretic perspective. Wiley Blackwell

6. Mueller A (2007) Efficient wireless mac protocols-analyzing quality of service in wireless networks. VDM Verlag Saarbrücken, Germany
7. Perkins C (2003) Ad hoc on-demand distance vector routing (RFC 3561). 37 pp. http://www.ietf.org/rfc/rfc3561.txt
8. Perkins C, Royer E (1999) Ad-hoc on-demand distance vector routing. IEEE workshop on mobile computing systems and applications, pp 90–100
9. Royer E, Perkins C (2000) An implementation study of the AODV routing protocol. IEEE Conf Wireless Commun Netw 3:1003–1008
10. Das S, Perkins C, Royer E (2000) Performance comparison of two on-demand routing protocols for ad hoc networks. IEEE INFOCOM 2000 1:3–12
11. Gwalani S, Belding-Royer E, Perkins C (2003) AODV-PA: AODV with path accumulation. IEEE Int Conf Commun 1:527–531
12. Qadri N, Liotta A (2009) Analysis of pervasive mobile ad hoc routing protocols. pervasive computing: innovations in intelligent multimedia and applications. Springer, Berlin, pp 433–453

Chapter 6
Proactive Networks

Abstract What is the secret of a fast-responding network? The ideal network will anticipate the communication needs of all nodes, building the necessary paths proactively. Unfortunately, this level of intelligence is not possible today. A brute-force approach, whereby the network continuously maintains all possible paths among all nodes, is also not a viable proposition because networks are far too vast and dynamic. This chapter explores strategies to reduce the impact of signaling in proactive routing. Through this exercise, we find that a fast network is one that can adaptively switch between "proactive" and "reactive" modes.

The more precisely the position is determined, the less precisely the momentum is known in this instant, and vice versa
W. Heisenberg, Uncertainty Paper, 1927

6.1 From Reactive to Responsive

What is the secret of a fast-responding network? In Chap. 5, we assessed one of the dimensions: the ability to build paths "when" and "where" they are most needed. On-demand protocols such as AODV [1, 2] show us how to implement a network that has two modes of operation: a "standby" mode, which consumes minimum energy; and a "path construction" mode, which creates the necessary communication channels. In this case, most of the resources are dedicated to "adapting" to the user's demand.

However, a network that is able to respond to external stimuli is not necessarily a fast-responding network. Another essential feature is "proactiveness," that is, the ability to anticipate the communication needs while providing readily available paths.

A. Liotta and G. Exarchakos, *Networks for Pervasive Services*,
Lecture Notes in Electrical Engineering, 92, DOI: 10.1007/978-94-007-1473-1_6,

Ideally, one would want a network whereby every node knows the path to every other one. Unfortunately, the networks of today (and even more so regarding future networks) are by far too vast, dynamic and dense for any protocol to be able to capture their global status. For instance, let us consider the Link State (LS) protocol (Chap. 2) which strives to keep every node informed about the whole network topology. It does so by continuously broadcasting LS packets. However, the LS flooding process itself alters the very same topological status that it is trying to capture—LS packets incur extra traffic and delay. Also, the information carried by the LS packets may become obsolete while it is propagating through the network.

In this chapter, we address the question of "how to build a fast-responding network" in two steps. First, we explore proactive protocols that can work on highly dynamic networks. The key challenge is to reduce the impact of signaling without downgrading route accuracy.

Through this exercise, we find that, just like for "mere reactiveness," mere "proactiveness" does not always lead to fast responses. In the second part, we show how hybrid protocols, which strike a good balance between reactive and proactive mechanisms, bring about the best features of the two.

6.2 Keep the Network Ready

A good network is a "ready" network; however, keeping the network in "ready" mode generates signaling costs. As the network grows and becomes more dense and dynamic, we must find ways to reduce the size and the number of the signaling packets. A fine example of how this can be achieved is offered by OLSR, the optimized link state routing protocol [3, 4]. In OLSR, only a subset of the nodes is allowed to declare and relay the *topology control* (TC) messages. Such nodes are termed multipoint relays (MPR). The other nodes can receive and process TC packets, though do not retransmit them.

In practical terms, this means that we contain the signaling packets within a subset of the network. For maximum efficiency, this control network (the one carrying the TC messages) should remain small even when the number of nodes increases. In this way, we only need to flood a subset of the network that we are trying to control.

It is equally important that the size of the TC messages remains small. If TC packets carry status information about all neighbors, the packet size grows with the network density. To address this issue, OLSR makes a clever selection of a subset of neighbors whose link status can be advertized by TC messages.

Over the past few years, researchers have studied the properties of OLSR, using this protocol as the basis for more elaborate solutions [5–7]. OLSR is appealing because it brings the stability properties of Link State routing into the mobile scenery. Thus, it is essential to examine how OLSR constructs the control sub-network [8, 9].

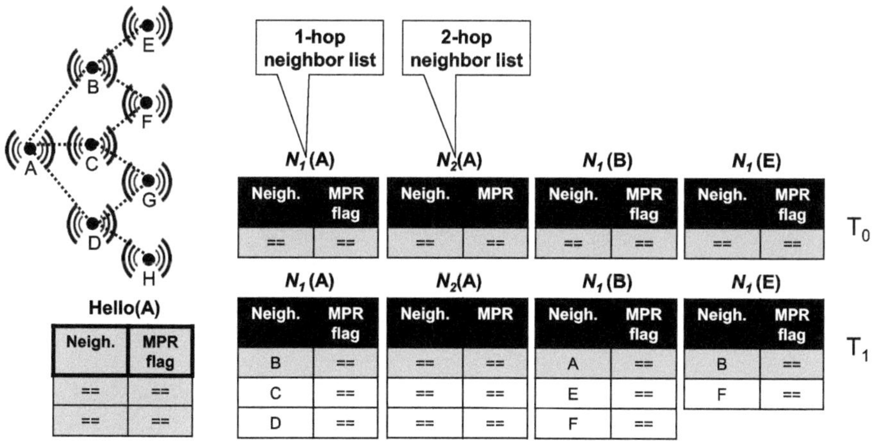

Fig. 6.1 Neighbor's discovery: periodic *Hello* messages

6.3 How Do I Find My Multipoint Relay?

Each node will periodically broadcast a tiny *Hello* packet which advertizes the node's neighbors. Thus, when a node captures a *Hello* packet, it actually discovers the neighbor plus the neighbor's neighbors. Take, for instance, the simple topology of Fig. 6.1, assuming that, initially, (time $T = T_0$) nobody knows about any other nodes. Hence, the list of 1-hop and of 2-hop neighbors (N_1 and N_2 respectively) will be empty; thus, the first set of *Hello* messages will carry no information.

At time T_1, node A learns about its neighbors B, C and D; so it can compile the 1-hop neighbor list, $N_1(A) \equiv \{[B, -]\ [C, -]\ [D, -]\}$. The MPR flag indicates whether or not some other node has chosen node A as its own multipoint relay node (this information is unknown at this stage).

However, at this stage, node A does not have sufficient information to fill in its 2-hop neighbor list, i.e., $N_2(A)$. At the following exchange of *Hello* packets, node A learns about E, F, G and H via the *Hello* packets received from B, C and D. There is now sufficient information to compile $N_2(A)$, as depicted in Fig. 6.2.

We still need to compute the MPR nodes of node A, that is, the minimum set of 1-hop neighbors through which A can reach all its 2-hop neighbors. At this point, node A has all the information required to complete such a task. We first establish whether a single 1-hop node can serve the purpose. We can easily see that B, C, nor D can individually reach E, F, G and H. Thus, we consider all possible combinations of pairs. The pair {B; C} cannot reach H. The pair {C; D} cannot reach E. However, {B; D} can reach all 2-hop nodes. Thus, B and D are elected as MPR nodes for node A, and the corresponding flags in $N_1(A)$ are set to "yes" accordingly (Fig. 6.3). The MPR nodes computed by each of the nodes are depicted in Fig. 6.4.

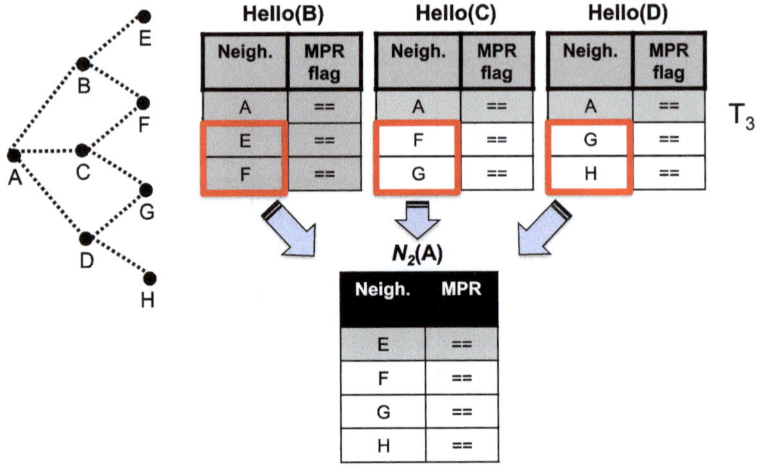

Fig. 6.2 Neighbor's discovery: the 2-hop neighbor list of node A, N_2(A)

Fig. 6.3 Computation of MPR(A), the minimal set of 1-hop neighbors that allows node A to reach all of its 2-hop neighbors

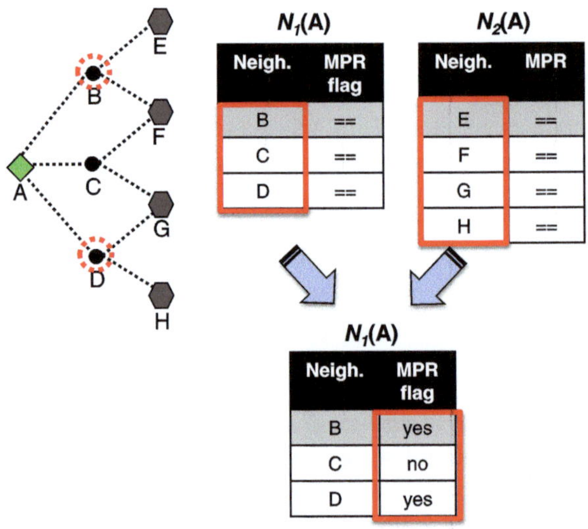

6.4 Life of an OLSR Node

Each node of an ad hoc network is a routing entity. Thus, every node must contribute to the tasks of *route computation* and *packet forwarding*. For this purpose, the nodes must execute different processes simultaneously, as delineated in Fig. 6.5. We have already discussed thread (a), i.e., the MPR selection. This process runs continuously because the node dynamics can lead to new configurations and, possibly, to new MPRs.

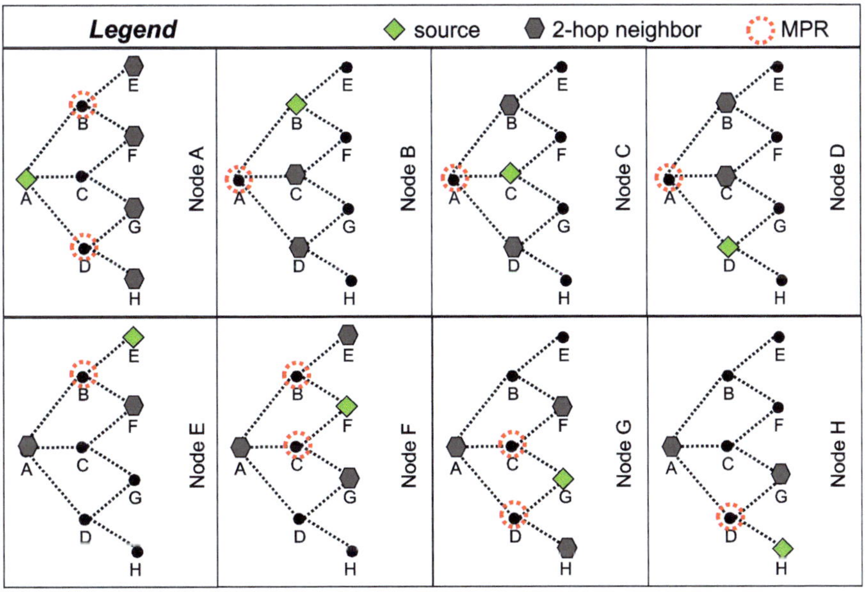

Fig. 6.4 The MPR nodes in our sample network

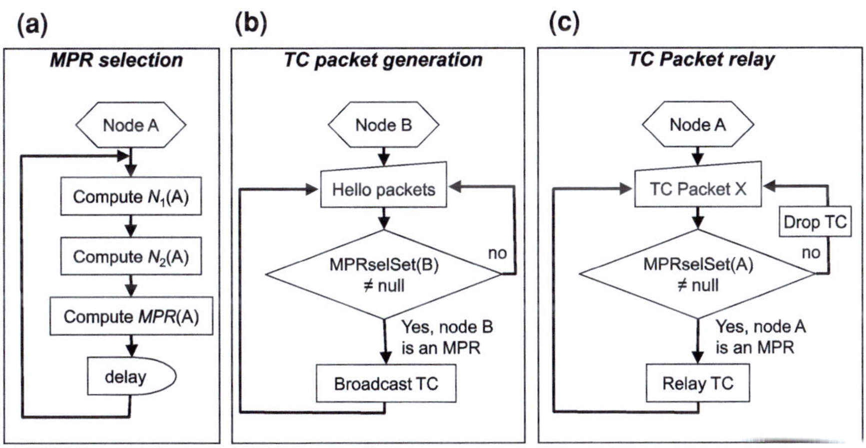

Fig. 6.5 The main threads of an OLSR node: **a** MPR selection, **b** TC packet generation and **c** TC packet relay

To limit the flooding process, only the MPR nodes are allowed to generate TC packets (Fig. 6.5b). However, how does a node learn whether or not it is the MPR for some other node? This information propagates thanks to the *Hello* messages (incorporated in the *MPR flag* field). For instance, node B (which is MPR of A) will periodically broadcast TC messages. Figure 6.6a shows how the various *Hello*

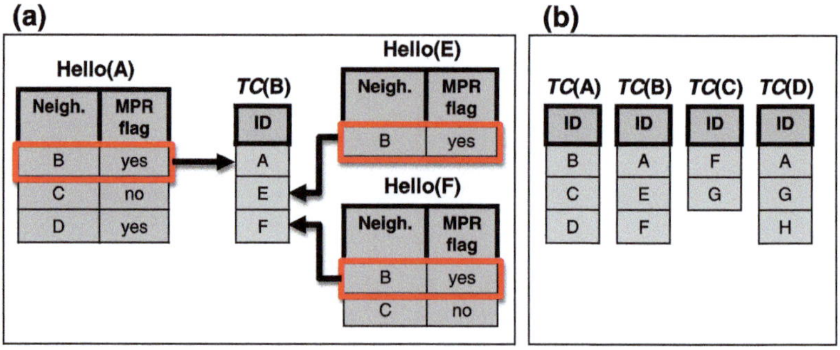

Fig. 6.6 **a** The MPR selector set of node B and its TC packet; **b** the other TC packets

messages received by B lead to the formation of its MPR selector set and, subsequently, to the construction and broadcasting of the TC packet. The other TC packets of this simple network are visible in Fig. 6.6b. Nodes E, F, G and H are not selected as MPR by any node. Therefore, they do not generate any TC message.

In addition to the processing of *Hello* packets and to the generation of its own TC packets, MPR nodes must also support the propagation of the other TC messages (Fig. 6.5c). The non-MPR nodes will simply drop the TC messages.

6.5 The Node's Information Repository

Thanks to the continuous exchange of *Hello* messages and TC packets, each node gradually builds up its own information repository. This includes the 1-hop neighbor list, the 2-hop neighbor list, and the MPR selector list. As new TC packets come in, the node will also start building the topology information base (TIB) that lays the foundations of the routing table. The construction of the TIB is sketched in Fig. 6.7a. As an example, Fig. 6.7b includes the complete information repository of node A.

6.6 Shortest Path over the MPR Sub-topology

At this point, each node has all the data necessary to compute the shortest paths. Let us continue to follow the process from the point of view of node A captured in Fig. 6.8. The first three entries of the routing table are directly derived from the 1-hop neighbor list, $N_1(A)$. Nodes B, C and D are neighbors, so they are directly visible by A (next-hop addresses are B, C and D respectively) and their distance from A is equal to one.

To compute the other entries, we can just apply the Dijkstra's shortest-path algorithm (see Chap. 2) on the MPR sub-topology of our network. Figure 6.8b

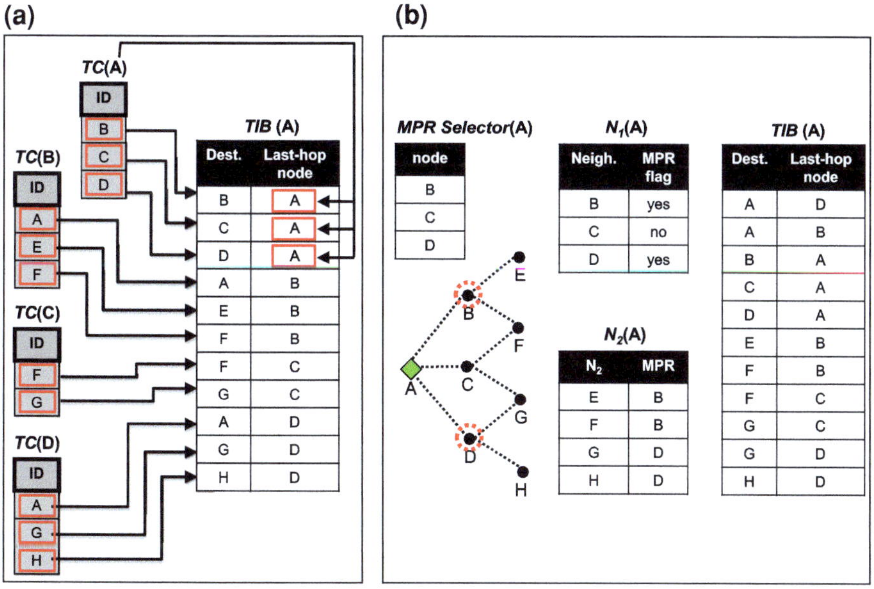

Fig. 6.7 **a** Computation of the topology information base (TIB) of node A. For every TC message received, its entries are copied in the first column of the TIB. The originator ID is copied onto the second column. In this way, the TIB records pairs of node IDs together with their respective 1-hop neighbors; **b** the complete information repository of node A

depicts the calculation of the routing entry for destination E. We visit the TIB table to find out that one neighbor of node E is node B. Then, we continue the same process recursively until we find a *destination* entry whose *last-hop* node is node A. In this simple case, the algorithm converges in only two steps.

A slightly more complex case is depicted in Fig. 6.8c. We are now looking for the path A → F of which there are two alternatives. We follow either B or C. However, if we look at the list of MPR, we find that node C is not included. Thus, the only acceptable path (i.e., via an MPR node) is the one via node B. After a second look up of the TIB (looking for the *last-hop* for B), the algorithm converges and we have found the routing entry {F; B; 2}.

The computational process performed to obtain routes for destinations G and H follows exactly the same steps as above, and eventually leads to the shortest-path distribution tree of Fig. 6.9.

6.7 A Complete Example

Figure 6.10 presents a complete OLSR network example, including the main intermediate steps through which node X builds its path to B. The first task is to identify the leaf nodes within two hops of the source node (X) and find the set of 1-

(a)

N_1(A)

Neigh.	MPR flag
B	yes
C	no
D	yes

ROUTING TABLE of A

Dest	Next	Dist.
A	==	==
B	B	1
C	C	1
D	D	1

(b)

TIB (A)

Dest.	Last-hop node
A	D
A	B
B	A
C	A
D	A
E	B
F	B
F	C
G	C
G	D
H	D

TIB (A)

Dest.	Last-hop node
A	D
A	B
B	A
C	A
D	A
E	B
F	B
F	C
G	C
G	D
H	D

ROUTING TABLE of A

Dest	Next	Dist.
A	==	==
B	B	1
C	C	1
D	D	1
E	B	2
F	B	2
G	D	2
H	D	2

(c)

TIB (A)

Dest.	Last-hop node
A	D
A	B
B	A
C	A
D	A
E	B
F	B
F	C
G	C
G	D
H	D

TIB (A)

Dest.	Last-hop node
A	D
A	B
B	A
C	A
D	A
E	B
F	B
F	C
G	C
G	D
H	D

ROUTING TABLE of A

Dest	Next	Dist.
A	==	==
B	B	1
C	C	1
D	D	1
E	B	2
F	B	2
G	D	2
H	D	2

Fig. 6.8 Shortest-path calculation in OLSR. **a** The first entries of the routing table are derived directly from the 1-hop neighbor list; **b** routing entry for destination node E; **c** routing entry for destination node F

hop nodes that covers those leaves (a). Node Y is the only leaf and is covered by node S which is thus elected as the first MPR (b). Since we now have an MPR, let us see which 2-hop nodes it is covering (c). Usually, the process would continue until a minimum set of MPR that covers all 2-hop nodes is established. However, in our case, node S covers all 2-hop nodes. Thus, we can already draw the initial part of the distribution trees rooted at X (d).

The subsequent part of the tree is built by the MPRs. In Fig. 6.10e, we see the resulting distribution tree routed at node S, i.e., the MPR of node X, now operating in the role of source node. In (f), we see how the path between X and B starts taking form once the routing tree of node X is merged with that of node S

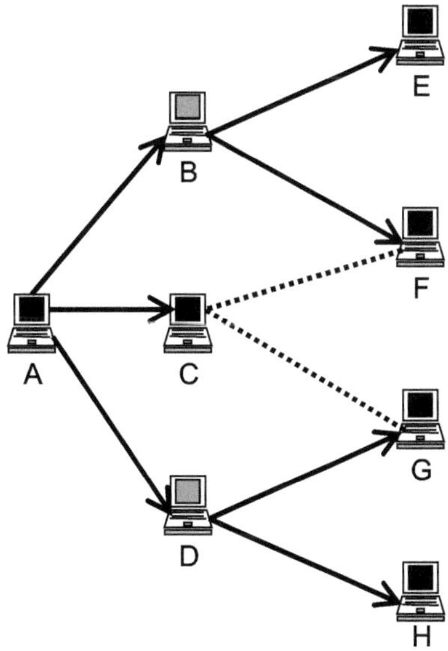

Fig. 6.9 Shortest-path calculation distribution tree rooted at node A and built over the MPR sub-topology

(for simplicity, we only show the paths leading towards B). In (g), we can observe a portion of the routing tree of node M (i.e., MPR node of S and now acting as a source node). Finally, the complete path is shown in (h). Furthermore, the complete distribution tree routed at node X is depicted in Fig. 6.11b The new versions of OLSR allow a more efficient distribution of control information, based on the Fisheye routing algorithm [10].

6.8 How Proactive Can You Be?

Proactive protocols such as OLSR strive to keep the network ready at whole times, furnishing it with routing tables. However, proactiveness is effective only when the tables reflect the actual status of the network. At the same time, this table-driven approach comes with signaling overheads that affect the very same network (and tables) they are trying to control. Thus, when the control traffic starts occupying a significant portion of the available network capacity, the costs incurred to build the tables outweighs the benefits of proactive routings.

OLSR offers a strategy to reduce the control traffic by containing it within the MPR sub-topology. This is effective in "dense" networks because, in this case, each MPR node serves many neighbors and the sub-topology stays considerably smaller than the whole network. Thus, a great portion of the nodes and links is not affected by the signals. On the other hand, if the nodes are sparsely distributed, the benefits of OLSR in comparison to the ordinary link state protocol quickly vanish.

The level of network dynamics is one of the factors that determine whether or not a proactive approach is better than a reactive one. Clearly, if the nodes move

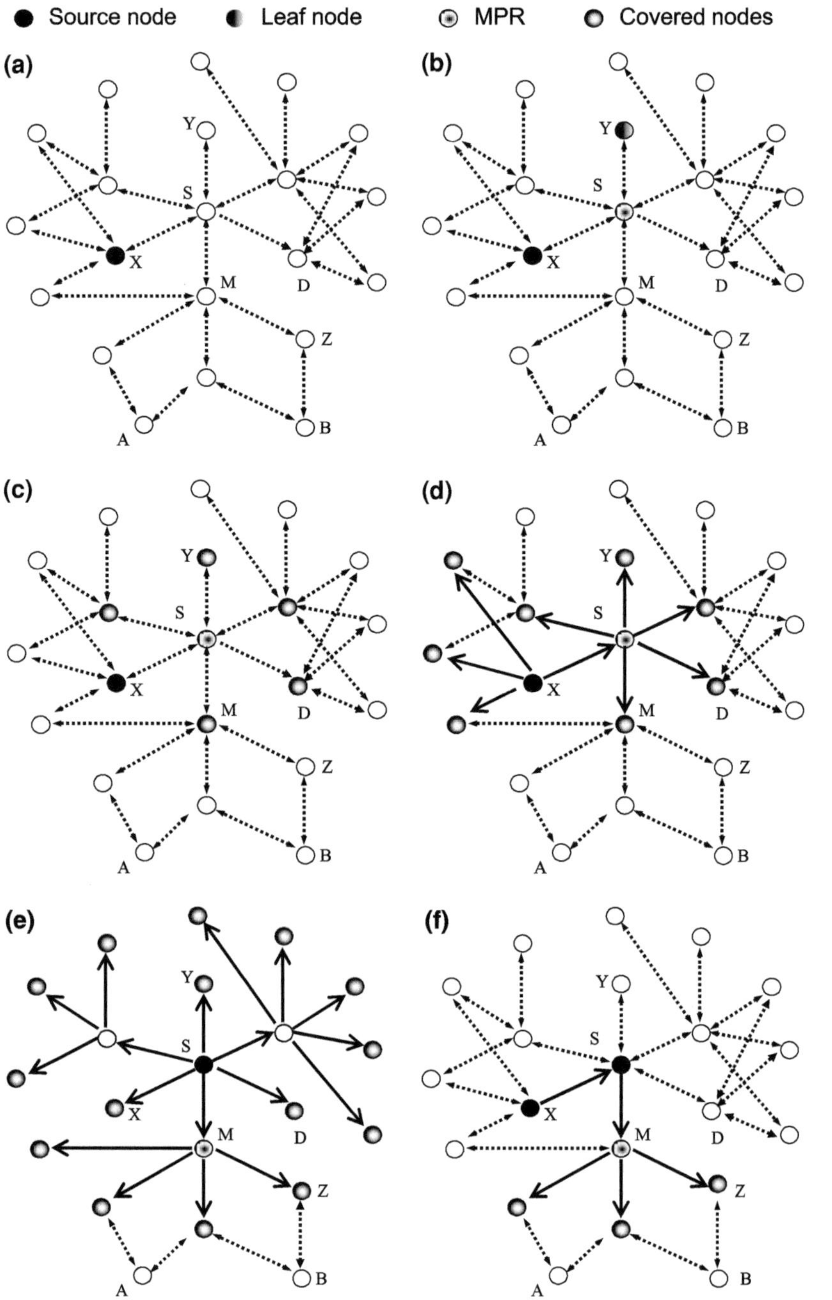

Fig. 6.10 **a–h** Shortest-path calculation between nodes X and B

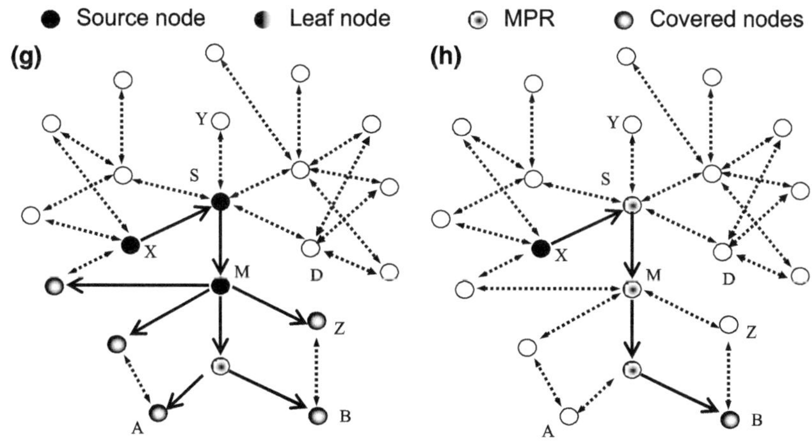

Fig. 6.10 (Continued)

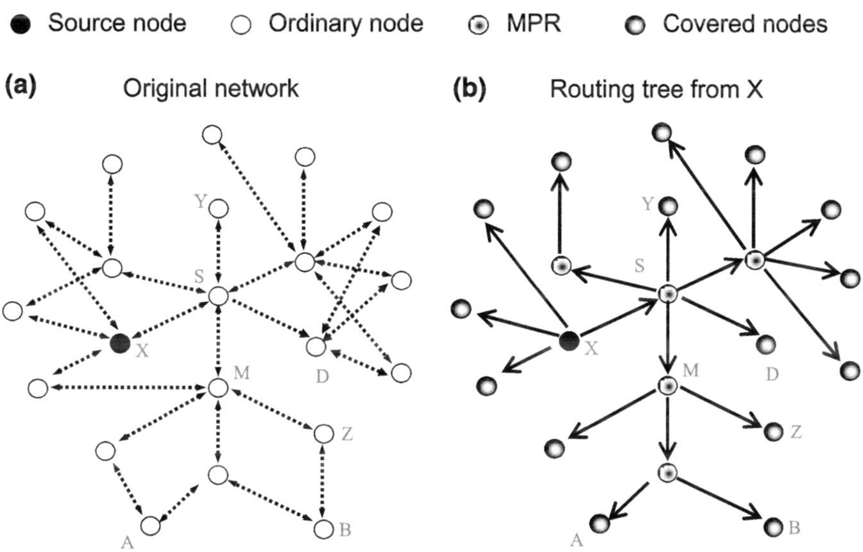

Fig. 6.11 OLSR distribution tree. **a** Initial network topology; **b** distribution tree rooted at node X

too fast, the control packets carry outdated information; thus, the routing tables are permanently out of sync with the network status. In this case, the whole exercise of maintaining tables becomes useless, which sheds a positive light on reactive routing. Together with node mobility, link failure tends to increase network dynamics. Thus, failure-prone networks demand a good degree of reactiveness.

Another important factor is the typology of the data traffic. If the traffic sources and destinations only involve a small subset of nodes, reactive routing represents a more efficient strategy: all the control power is diverted towards the active paths

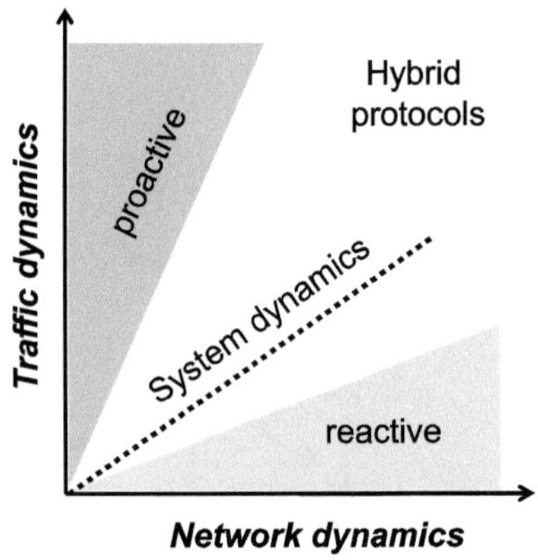

Fig. 6.12 System dynamics and the role of proactive, reactive and hybrid protocols

and no energy is dispersed in the calculation of the other unutilized ones. On the same end of the scale, we have the self-similar traffic: there is a repetitive pattern among the inter-communicating points. Therefore, the reactive routes are automatically kept alive by the traffic. On the other end, when the traffic is unpredictable and hits a larger fraction of the network, proactive protocols win.

Application patterns and network dynamics are relative, rather than absolute, metrics. In fact, the overall level of dynamics of the system depends on how the traffic dynamics compare with the network dynamics. Figure 6.12 projects a qualitative view on the relationship between the system dynamics and the choice of routing protocols. We can see that there is a big gap between proactive and reactive protocols. There is a broad range of conditions for which neither approach functions at its best. This is where hybrid solutions find their role in networking.

6.9 The Power of Hybrid Protocols

It is not uncommon that different portions of the network exhibit different levels of dynamics. Sometimes, we observe that most of the traffic moves within clusters, with occasional communications in between clusters. Also, different groups of nodes may have different mobility patterns. Thus, it makes sense to deploy both reactive and proactive mechanisms on the same network, and then switch between them depending on the localized network conditions. Hybrid protocols incorporate this mixture of capabilities and aim to combine the merits of reactive and proactive protocols. For instance, in the presence of traffic clustering, it makes sense to use proactive algorithms for the computation of local routes and the reactive approach to build longer paths on demand.

Hybrid routing is exemplified by the zone routing protocol (ZRP) [11]. Each node N can define the hop-radius r of its own routing zone. It will then adopt a neighbor discovery protocol (NDP) to find all the nodes residing within its

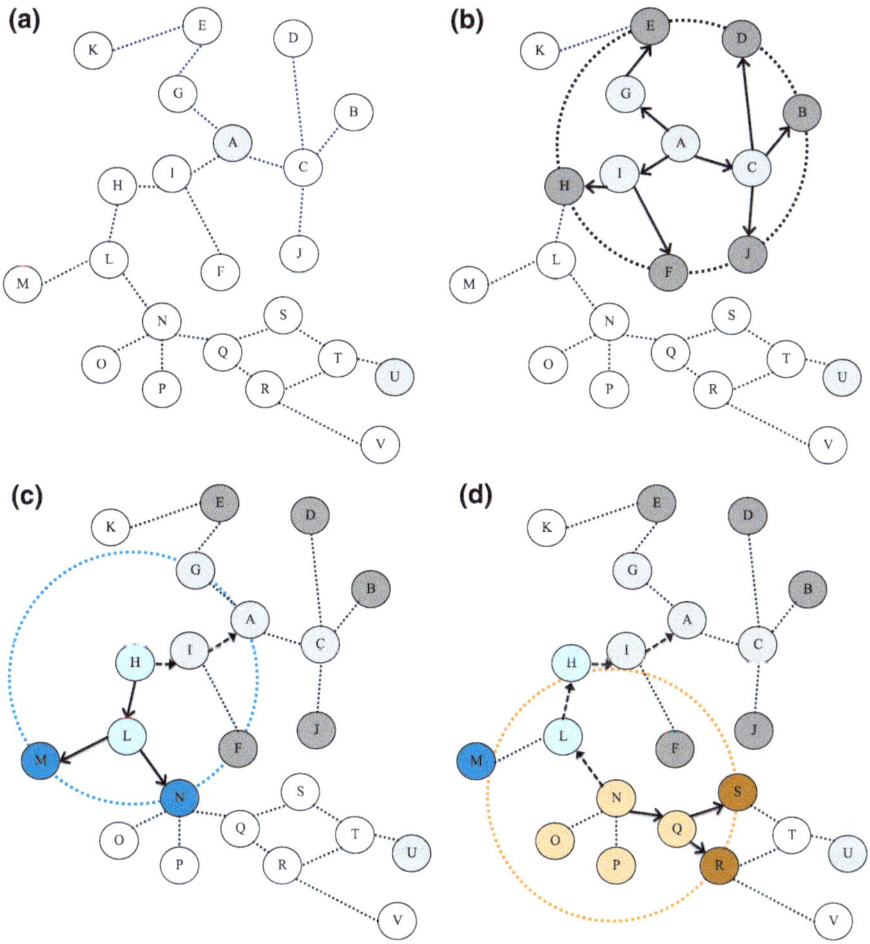

Fig. 6.13 a–f Discovery the of path from node A to node U according to the hybrid protocol ZRP

catching area. Nodes may join or leave the zone, but the node N will adopt a proactive routing protocol (such as OLSR) to maintain its in-zone distribution tree.

This operation is sufficient as far as N sends data within its own zone. On the other hand, as soon as one of the destination nodes falls outside of the zone, N triggers a reactive protocol (such as AODV) to build the missing part of the path from the border onwards.

Let us see which routes are computed proactively and which ones are build on-demand, considering the example of Fig. 6.13a. Let us start from node A and find out how it will eventually communicate with node U. Just like any other node, node A maintains its in-zone routes proactively. Thus, node A can only reach as far as its border nodes (B, D, E, F, J and H) (Fig. 6.13b).

Because node U does not reside within the zone, node A has to initiate a reactive search through its boarder nodes. This process is known as "bordercast" (not to be

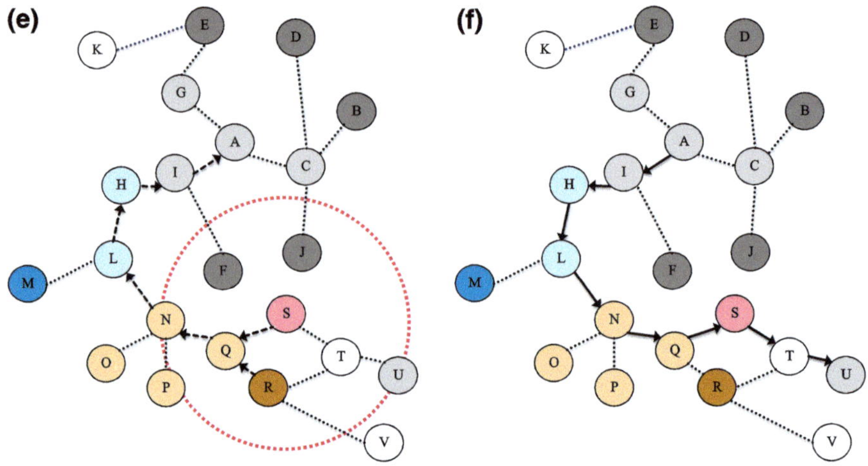

Fig. 6.13 (Continued)

confused with "broadcast") because the route discovery request is received and processed only by the border nodes. Let us follow how this request is executed via node H—the other border nodes will follow a similar procedure but will fail to discover node U. Suppose we are adopting the reactive AODV protocol (Chap. 5) to discover the paths in between zones. Node H builds a reverse path (H → I → A) that will eventually be used by node U to send route reply (RREP) packets back to A (Fig. 6.13c). Node H has its own zone as well; therefore, it can immediately provide proactive paths to its boarder nodes (M, N and F). Since destination node U is, again, not found, node H issues a reactive bordercast request.

Let us follow how this request is executed via node N. The process is the same as the one followed by node H. Node N builds the reverse path (N → L → H) and provides another bit of the path leading to the boarder (Fig. 6.13d). It will then boardercast a route discovery request via nodes R and S.

We are now getting closer to our destination; in fact, both R and S have a valid path to node U. Suppose that node S is the first who responds: it builds the reverse path (S → Q → N), but this time there is no need to invoke AODV since node S already holds a proactive path to node U (Fig. 6.13e). Finally, node S sends the RREP back to A via the reverse path and the direct path is set up (Fig. 6.13f).

In ZRP, each node can adjust its own radius depending on the level of dynamics experienced within the zone. When the node's mobility is low, the routes tend to be more stable. The node can thus increase the radius of the zone with little or no effects on the accuracy of the routes. On the other hand, as the mobility increases, the routes break more frequently and the cost of maintaining proactive paths quickly becomes prohibitive. Hence, as the zone dynamics increases, it is more convenient to decrease the radius.

Hybrid protocols offer extra means to increase the network's ability to adapt to the application. We can say that protocols such as ZRP come with two gears (the "proactive" and the "reactive" modes). However, networks with multiple gears involve a more complex orchestration of the system's parameters. Each node must be able to self-regulate its own radius, but also dynamically adjust all the parameters of both its intrazone routing protocol (IARP), e.g., OLSR—and interzone routing protocol (IERP), e.g., AODV. Thus, self-adaptability comes with an additional challenge: to furnish the nodes with the ability to gather their context, reason about it and adjust the various network parameters coherently. The ultimate self-adjustable network will pursue maximum efficiency whilst simultaneously avoiding deadlocks, loops, local sub-optimality and instability.

We came to explore the meanders of ad hoc networks, seeking inspiration for the quest of the future Internet. Ad hoc networks represent a unique experimental ground for the study of proactive, reactive and adaptive protocols. In fact, as networks become increasingly dynamic, transient, volatile and error-prone, we must equip them with effective self-healing and self-optimization mechanisms [12]. In this sense, the scientific framework developed in the context of ad hoc networks has generated ideas and protocols that will definitely play a role in the future.

On the other hand, the recent advances on "virtual networks" pursue a completely different perspective. Starting from the next chapter, we shall explore how to build a new breed of networks on top of a general-purpose communication system.

References

1. Perkins C (2003) Ad hoc on-demand distance vector routing (RFC 3561). 37 pp. http://www.ietf.org/rfc/rfc3561.txt
2. Perkins C, Royer E (1999) Ad-hoc on-demand distance vector routing. In: IEEE workshop on mobile computing systems and applications, pp 90–100
3. Clausen T, Jacquet P (2003) Optimized link state routing protocol (RFC 3626). 75 pp. http://www.ietf.org/rfc/rfc3626.txt. Accessed 7 Feb 2011
4. Clausen T, Jacquet P, Adjih C, Laouiti A, Minet P, Muhlethaler P, Qayyum A, Viennot L (2003) Optimized link state routing protocol (INRIA report 5145)
5. Haas Z (1997) A new routing protocol for the reconfigurable wireless networks. In: IEEE international conference on universal personal communications record, vol 2, pp 562–566
6. Mian A, Baldoni R, Beraldi R (2009) A survey of service discovery protocols in multihop mobile ad hoc networks. Pervasive Comput IEEE 8:66–74
7. Qadri N, Liotta A (2009) Analysis of pervasive mobile ad hoc routing protocols. Pervasive computing innovations in intelligent multimedia and applications. Springer, Berlin, pp 433–453
8. Sarkar SK, Basavaraju TG, Puttamadappa C (2007) Ad hoc mobile wireless networks. CRC Press, Boca Raton
9. Tonguz O, Ferrari G (2006) Ad hoc wireless networks: a communication-theoretic perspective. Wiley, Hoboken
10. Pei G, Gerla M, Chen T-W (2000) Fisheye state routing: a routing scheme for ad hoc wireless networks. IEEE Int Conf Commun 1:70–74
11. Haas Z, Pearlman M, Samar P (2002) The zone routing protocol (ZRP) for ad hoc networks. 11 pp. http://tools.ietf.org/id/draft-ietf-manet-zone-zrp-04.txt
12. Jennings B, van der Meer S, Balasubramaniam S, Botvich D, Foghlu M, Donnelly W, Strassner J (2007) Towards autonomic management of communications networks. Commun Mag IEEE 45:112–121

Chapter 7
Content-Aware Networks

Abstract Packet switching networks provide rudimentary means to move units of "raw" data around. The Net can "transport," though it is unable to "manipulate" high-level content, video or audio sessions. Imagine what we could achieve with a network that is redesigned around what is the most precious thing in today's digital ecosystem: *the content*. This chapter introduces content-aware networks, ones that can re-route packets based on the content "usage patterns" and "requirements." We look at peer-to-peer networks as a practical example in order to better understand how to build content-aware networks on top of ordinary packet switching networks.

With virtue you can't be entirely poor; without virtue you can't really be rich

Chinese proverb

7.1 Routers Should Read the Content

Packet switching networks deliver content quantized in blocks between any two addresses. This point-to-point delivery is simple and efficient, but is only a small subset of the transport services a network should provide. For instance, how could a node push content to other interested nodes without building a separate delivery session for each one of them? One-to-many and many-to-many distribution schemes are not fully supported by IP networks. These are purposely kept ignorant about the data plane (including the content features and distribution patterns) as well as the physical context where the information is delivered (including the physical network, the users' terminals and their location). Because of their isolation from the rest of the system, routers perform their tasks sub-optimally. They relay raw bytes and are not concerned with information and communication patterns.

A. Liotta and G. Exarchakos, *Networks for Pervasive Services*,
Lecture Notes in Electrical Engineering, 92, DOI: 10.1007/978-94-007-1473-1_7,

Imagine what we could achieve if routers were able to identify and understand the content, become aware of which nodes are interested in it and, then, create replicas accordingly. Not one of these steps is possible unless the network takes up tasks that are currently in the application domain.

In this chapter, we start a journey in the realm of content-aware networks, ones that can optimize content transfers based on the content usage patterns. Content-aware networks are made of "virtual" routing nodes in the sense that the routing functions are performed by software entities rather than by specialized (dedicated) hardware. These virtual nodes can physically intercommunicate via a general-purpose physical network; but they are also interconnected by "virtual" links that play a key role in content-aware routing.

To understand the mechanics of content-aware routing we look at protocols and mechanisms developed in the context of P2P networks [1–4]. In this case, the user's terminals, termed "peers," take up the "virtual routing" function and the overall routing protocol is implemented in the form of an application (the P2P program). P2P networks are optimized for specific tasks such as file-sharing, video on demand, IPTV or video conferencing. In consideration of their characteristics, P2P networks belong to the category of "virtual networks." These can deploy their own routing strategies without having to change any protocol on the IP network.

7.2 A Network on Top of the Physical Network

There is complementarity between the P2P routing algorithm (which is content driven) and the underlying packet switching routing (which is content-agnostic). In the example of Fig. 7.1, the two colleagues might receive George's messages (*msg*) via Antonio's computer—from a peer's perspective, messages travel in between peers over the virtual links. However, the actual delivery medium must always be a physical network—the packets are transported by the switched network.

Virtual networks aim to compensate the lack of flexibility and limited efficiency of the underlying network. Thus, the virtual links do not always map directly onto the physical links. For instance, the same physical link may sustain multiple virtual links (Fig. 7.2). Also, the various packets of the same message may follow different paths (Fig. 7.3). In fact, as the network conditions vary (e.g., due to congestion), the routing protocol can create diversions.

Virtual networks go far beyond this dichotomy between virtual messages and physical packets. They generate an entirely new network on top of the physical network (Fig. 7.4). The peer nodes of the virtual network are chosen among those that have some form of intercommunication patterns. Each peer has a syntactic (sometime even a semantic) understanding of the message; therefore, it can more easily optimize the distribution of the message to the other peers. In fact, peers can advertize their content, discover the adverts of other peers and use this information to construct the virtual network. Thus, the virtual paths only exist as far as there is shared content among the nodes. This is very different from the infrastructure-

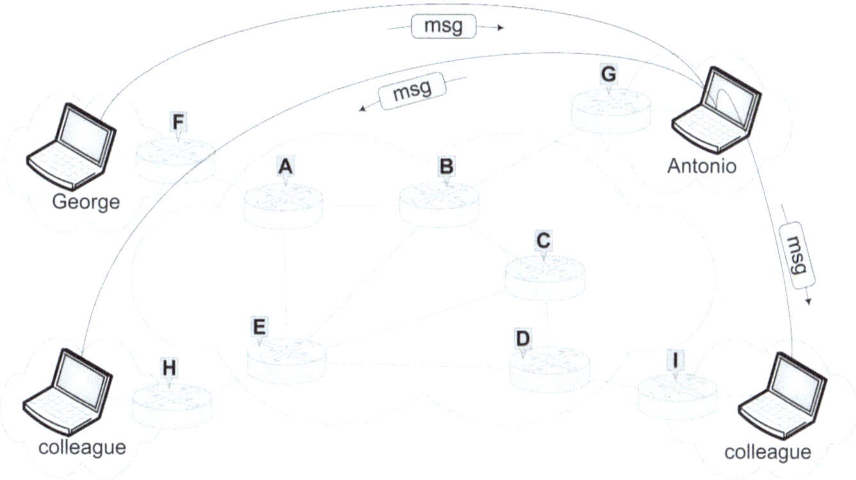

Fig. 7.1 Virtual links between peers at the edges of a packet switching network

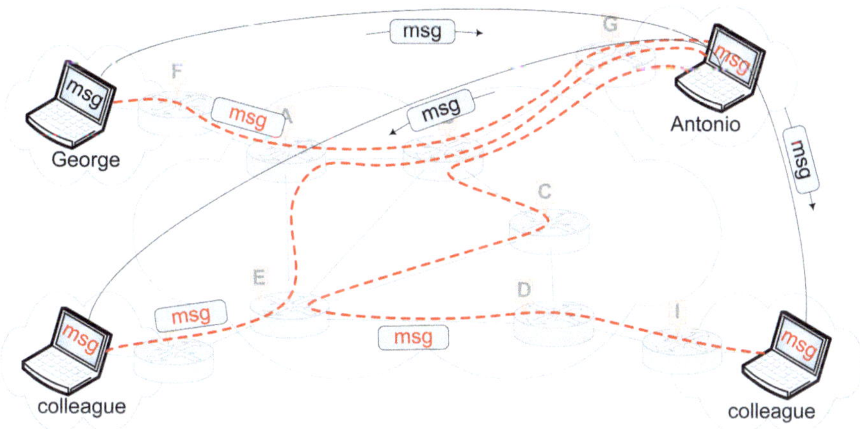

Fig. 7.2 Virtual links consist of a succession of routers and physical links

based networks which strive to keep everything connected. On the other hand, there are similarities with the reactive routing protocols (Chap. 5) which are also maintaining paths on the bases of packet transfers. However, in the ad hoc reactive networks, the network is still content-agnostic.

The nodes of virtual P2P networks can send, receive and relay information, thus acting in the roles of client, server and router. At its essence, a P2P network needs to offer mechanisms for nodes to join in, obtain a globally valid identifier (ID) and maintain a neighbor list. A peer will promote any of the other nodes to the status of "neighbor" when it recognizes some affinity in terms of shared content.

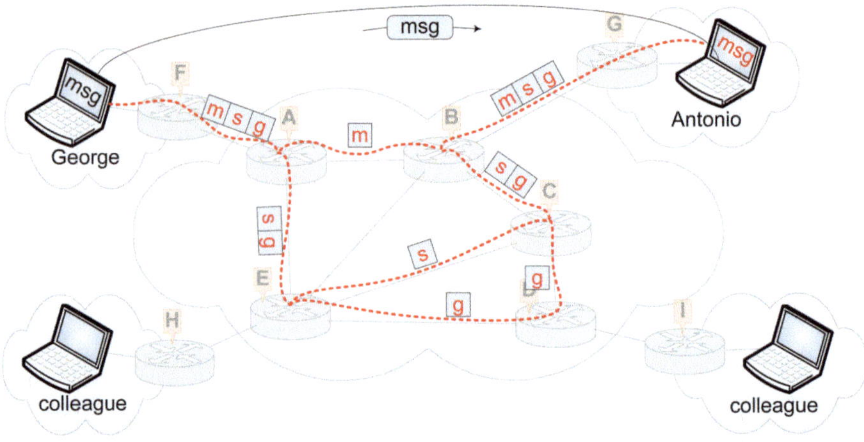

Fig. 7.3 Every virtual link (e.g., George–Antonio) may consist of multiple parallel paths

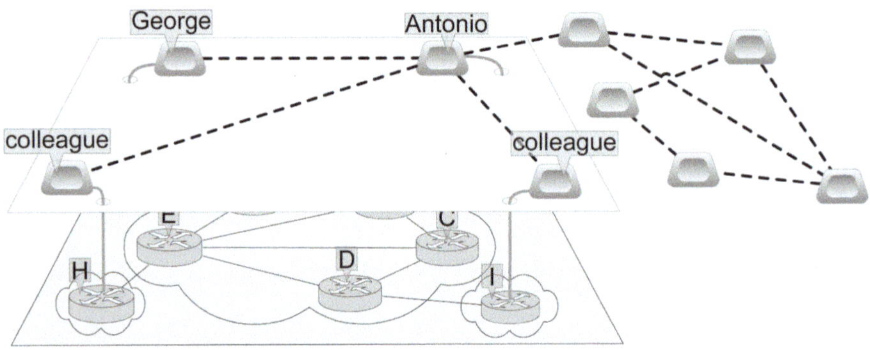

Fig. 7.4 Virtual network built with peers at the edges of a packet switching network

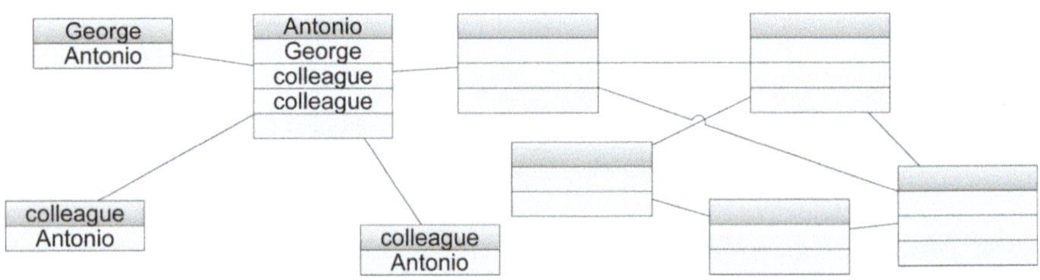

Fig. 7.5 Application-specific virtual networks build graphs of distributed neighbor lists

Hence, from a more practical point of view, a P2P network can be depicted as a distributed list of {ID, IP} pairs with redundancy. Each node maintains a portion of the neighbor list in order to keep relevant nodes in direct contact (Fig. 7.5). For the purpose of efficiency, we need to maintain the neighbor list small; so there is no point in maintaining direct links between nodes who share nothing in common.

7.3 Centralized Assignment of Node Identifiers

How does a P2P network bootstrap? Nodes will join the network one after the other. The first thing a new node should do is acquiring a unique identity (ID) within the virtual network [5]. Next, the node must find an entry point, i.e., any other node who is already into the network or knows how to get in. The joining process then continues with the population of the neighbor list. Each node augments its neighbor list with multiple entries and advertises its own resources.

The ID assignment is an important step of the bootstrapping process [6–8]. As with IP addresses in packet switching networks, node IDs in virtual networks may be static or dynamic. A static ID identifies a node throughout its online or offline lifetime; whereas a dynamic one may be different each time the node joins the same network. Static IDs are useful to nodes with a fixed set of neighbors—any node re-joining the network expects to virtually connect to the same static neighborhood. For instance, a Skype [9] account has a list of contacts with which the client tries to connect each time it goes online. On the other hand, a node with a dynamic ID does not have a predefined fixed neighborhood and requires a less complex management mechanism.

What if two nodes within a network have the same ID? Messages targeted to one node may end up with the wrong one, causing serious coordination or even security problems. A typical solution is to use another monitoring entity—the *bootstrapping server*—that has a global view of the network [10, 11]. The server issues unique IDs to the new nodes and keeps track of those who are online (Fig. 7.6). In case of static neighborhoods, the server makes sure that no two nodes are assigned to the same ID, regardless of whether they are online or offline (Fig. 7.6a). Note the sequence of events. We have two nodes, Antonio and George. At time T = (t−2), Antonio registers his IP address, obtaining the identifier A. He will then abandon the network at time T = (t−1). When Antonio rejoins the network, he gets a new IP address from the physical network. Yet, within the virtual network, he is still associated with the same identifier A. This causes some inconsistency at time T = t because this new pair {A; 238.19.7.81} is temporarily out of sync with the node G (G still thinks that identifier A is coupled with IP address 224.5.45.8).

Figure 7.6b depicts the case of dynamic ID assignment. Now, we can only ensure unique IDs among the pool of nodes who are online. Thus, the IDs of the offline nodes are recycled as new nodes come online.

7.4 Centralized Entry Point Discovery

Getting an ID is only the first step of a node's bootstrapping process. The new node needs to populate its neighbor list with the support of some other online node. Thus, the second bootstrapping step is the identification of an entry point, a node

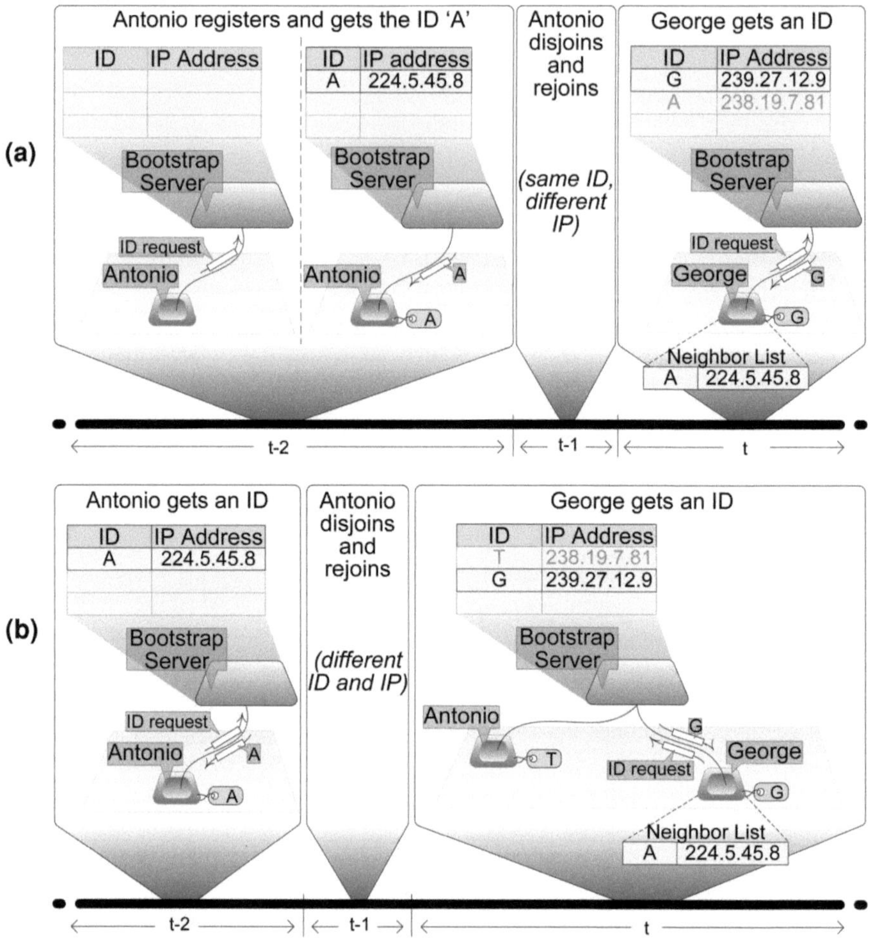

Fig. 7.6 Node ID issued from a well-known centralized entity having a global view of the system. **a** Nodes preserve the same ID either online or offline. **b** Nodes may get different IDs each time they rejoin

that will act as a gateway into the virtual network. For this purpose, many overlay networks commission an external entity, i.e., the bootstrap server [10]. This is in charge of monitoring the status of the virtual nodes to maintain two lists: the online and the offline nodes.

How does the server know about the status of the virtual nodes, considering that these may join in and out of the network in any moment? A simple approach is based on periodic *heartbeat* initiated by the nodes. When a node retrieves its ID from the server, it also finds out about the monitoring protocol in use and registers its presence with the server.

Even after registering with the server, a node cannot do much until it has discovered an entry point. This is a node which is already part of the virtual

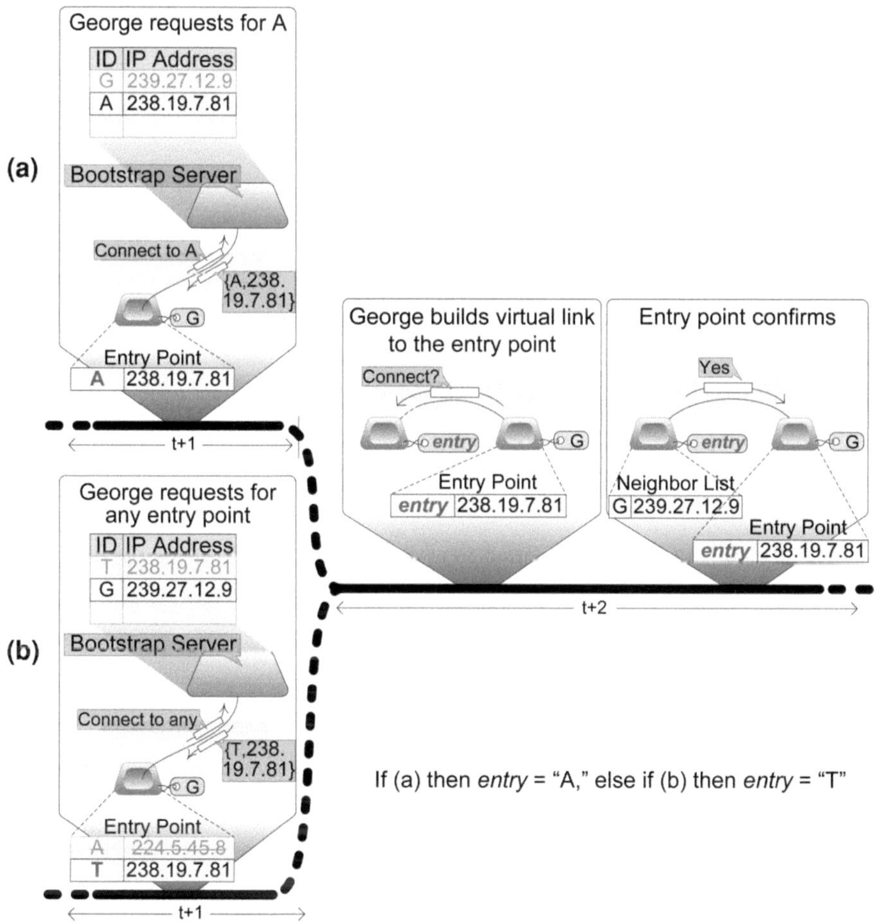

Fig. 7.7 Node G discovers its entry point via the bootstrap server. **a** If node G has fixed neighbors, it tries to retrieve their current IP addresses; **b** otherwise, it requests for any currently online {ID; IP} pair. Then, it connects and communicates with the retrieved entry point

network and will support the new comer in different ways. Continuing from the example of Fig. 7.6, let us see how node G will discover its entry point. In case of a fixed neighbor list (Fig. 7.7a), node G requests the {ID; IP} pair of a specific node—in our case, this is node A. From this point onwards, nodes A and G create a virtual link and can support each other.

In case of dynamic neighbor lists, as a new node gets online, it seeks the support of any entry point (not merely the entry points discovered previously). In fact, George will find out that Antonio now maps onto {T; 238.19.7.81} and sets this pair as the entry point (Fig. 7.7b).

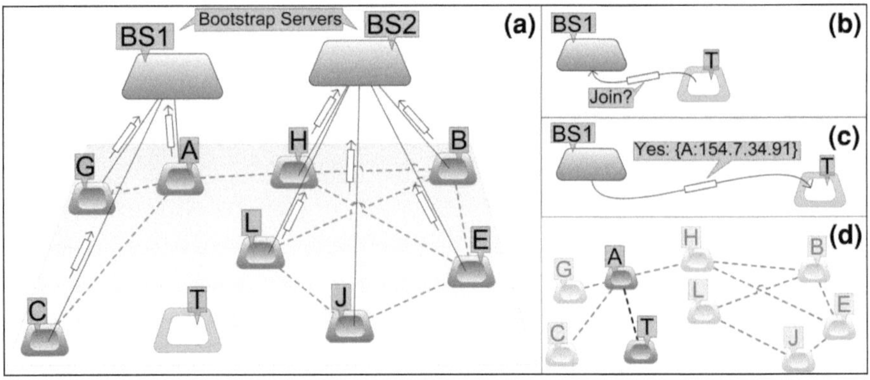

Fig. 7.8 Multiple independent bootstrap servers for one network. **a** Node T randomly picks one server among its preconfigured list of servers. **b** We suppose bootstrap server BS1 is selected. **c** BS1 returns entry point A. **d** Finally, the virtual link T to A is created

7.5 Multiple Bootstrap Servers

We have seen how the bootstrap server stays aware of the actual network and nodes status and fully controls the ID assignment/verification process. Bootstrapping is a vital service; therefore, if we rely on a single server, we are introducing a bottleneck and a single point of failure. To counter these limitations, many virtual networks use a farm of dedicated servers for a more robust bootstrapping process.

Examples of P2P systems that employ multiple servers are Skype, Yahoo, Windows Live messengers, Joost and eMule. In these cases, the peers come with a pre-configured list of possible bootstrapping servers. In some cases, the manager of these dedicated servers is also the owner of the virtual network.

The use of multiple servers allows balancing of the workload incurred by the joining nodes. Depending on the particular implementation, each server may either handle a subset of the nodes or hold a complete replica of the index. However, the use of servers implies that there is a management system behind the scenes of the virtual network. The servers' administrator has the capability to trace the users and may be held responsible for their actions, e.g., copyright infringement. In fact, this is how the music industry managed to shut down the file-sharing network Napster in 2001.

To avoid these vulnerabilities, some systems decouple the bootstrapping servers from the virtual network and decentralize their management [12, 13]. Let us assume that there are multiple such servers with no direct communication between them. Each server only has a partial view of the network; it only registers a subset of nodes. Figure 7.8 illustrates the process as node T joins the network and obtains the entry point via one of the servers.

As the virtual network increases in size and dynamics, more bootstrap servers may be added to decongest the existing servers. However, to further scale up the

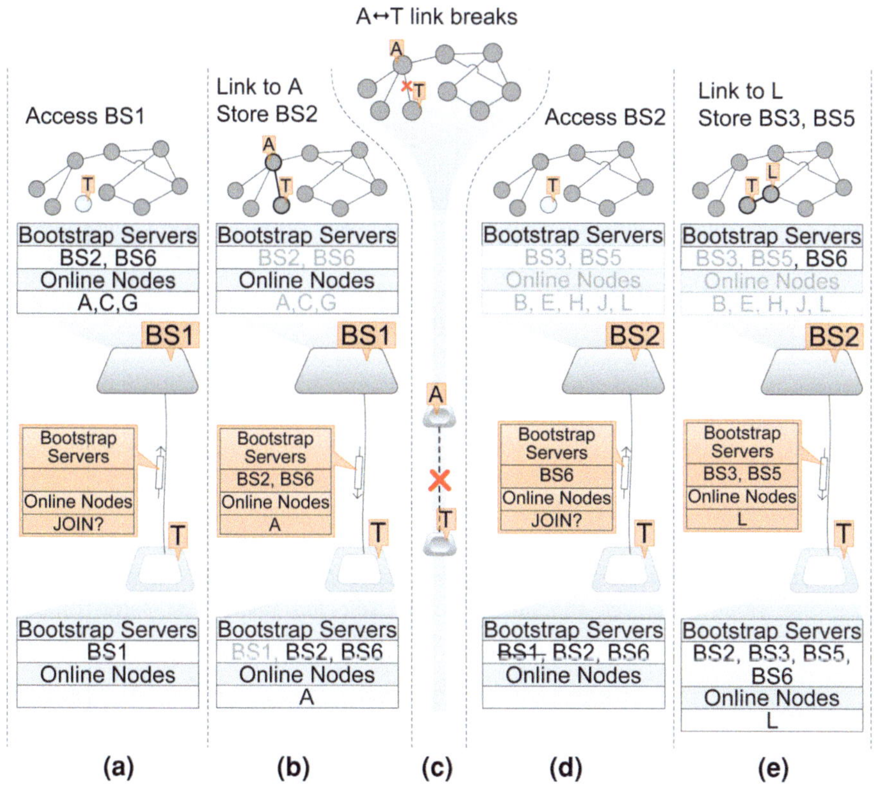

Fig. 7.9 Bootstrapping with the help of multiple independent servers. **a** Server BS1 is online; node T starts joining process. **b** Upon registering, node T receives entry point A as well as the list of other bootstrap servers (BS2 and BS6). **c** Both BS1 and node A go offline. **d** BS1 stays offline; node A comes online again but cannot re-register via BS1; node A registers via BS2 and at the same time propagates its own knowledge about the existence of BS6 to BS2. **e** Through the registration process, node T finds out about BS3 and BS5

system, at some point it is necessary to increase the level of redundancy and parallelism beyond the capability of a single server-based architecture. Figure 7.9 illustrates a scenario in which the actual peers participate to the bootstrapping process, propagating server availability information.

Following this approach, both the servers and the nodes help to keep the servers' list up to date. This scheme can take up substantial loads, thought it is still vulnerable because the virtual network still relies on the servers. Ideally, one would want the virtual network to function independently from any external entity in order to prevent issues on the servers from affecting the network. Even with multiple servers (instead of a single bootstrapping server), we still have scalability issues as these would still not be able to handle flash crowds, which are rather typical in P2P applications.

7.6 Decentralized Assignment of Node Identifiers

The bootstrapping process can also be performed in a fully decentralized fashion, i.e., no server involvement. Any node can initiate the generation of their own ID. They will use a hash function in combination with some local data to generate a unique ID. The local data may be the node's IP address, a local binary file or even a large random number. This is a probabilistic approach that cannot ensure the absence of ID duplicates; yet the probability that duplicates occur can be reduced by hashing onto a broad range of numbers.

While the centralized bootstrapping process has the benefit of increasing the sense of trust with respect to the network, the decentralized approach makes the process more robust and scalable. A third category of ID generation mechanisms tries to combine these two methods [6, 14, 15]. A new joining node produces only part of his ID, whilst some other nodes help completing the process. Following the example of Fig. 7.10, the new node T produces a random key which is then passed on to another randomly chosen node—in our example of Fig. 7.10a, this happens to be node A. In turn, node A combines the received key with another random locally generated number (the seed) and pushes the resulting value onto yet another randomly selected node. The same process continues, hopping from node to node, until the counter *hops-to-live* (HTL) has been decremented down to zero (Fig. 7.10b–d). Finally, the resulting ID is announced to the nodes that contributed to its generation by traveling backwards on the same path (Fig. 7.10e–g).

7.7 Entry Point Discovery via Underlying Links

In order to go through a decentralized bootstrapping process, a virtual node must be able to discover other nodes without any server support. Any new-coming node is initially isolated from the other virtual nodes. Its only door into the virtual network goes through the underlying physical network.

The new node has no prior knowledge. It has an IP address so it can only start by scanning the underlying IP network. On the other hand, the existing virtual nodes will be operating within the virtual network, but will also be listening on their IP address (on a specific port) for any incoming *join* requests. Thus, if a node *hears* a join request, it will respond (just as a bootstrap server would have done), becoming the entry point for the new node.

An example is depicted in Fig. 7.11. The incoming node, George, has obtained an IP address and it can start off by scanning its domain. Under these conditions, the scan is performed randomly. George generates a random IP (first within its domain). Eventually, if the IP address of virtual node Emily is selected, Emily will respond, becoming George's entry point. Otherwise, nodes from other domains might serve the same purpose.

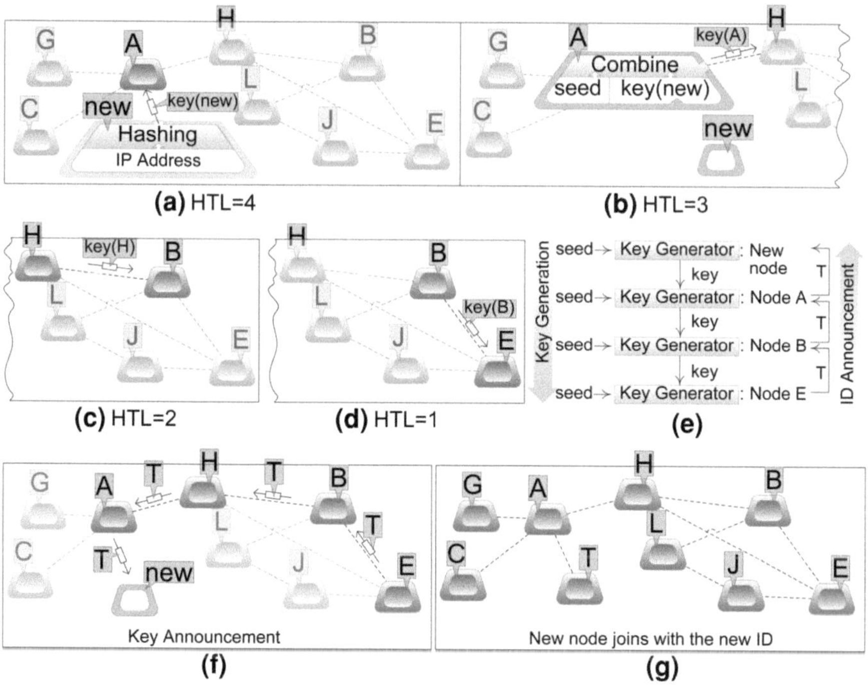

Fig. 7.10 Collaborative ID generation: **a–d** each node in a random path of HTL length generates a seed, combines it with the key received from the previous node and produces a new key to be passed onto the next. **f** The generated ID, T, travels all the way back to the new node. **g** Node T now has a unique identity within the network

This approach is merely a random probabilistic search over a *unicast* network. Its efficiency depends upon the ratio between virtual and physical nodes. Thus, the bigger the virtual network, the easier it will be for new nodes to join it. However, network scanning can be bandwidth consuming and does not give any guaranties as to the convergence time and success.

The entry point discovery process becomes more efficient if we can rely on additional support from the underlying network. For instance, if the network supports *multicasting* [16–18], we could map the virtual network onto a particular multicast group, i.e., onto a single multicast address. Thus, all virtual nodes would have the same network address. Each online node advertises its membership by registering the group address with the closest router (Fig. 7.12). In this way, the ordinary routing table updates have the effect to propagate the nodes' membership. For example, at some point, routers F, A and B become aware of Emily's membership; G, B and A register Antonio's presence. Thus, routers A and B enlist both Emily and Antonio as "reachable" on the same IP address, though via different paths.

It is up to the routers to forward the query to group members. For instance, George initially sends to router F a query with destination IP "230.70.70.70"

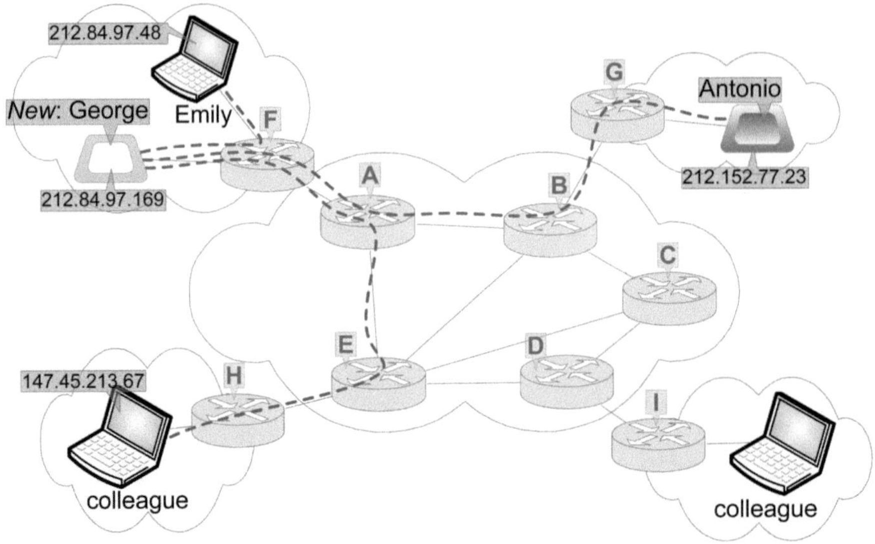

Fig. 7.11 Entry point discovery in a unicast network. George probes random IP addresses to discover a terminal hosting an online node

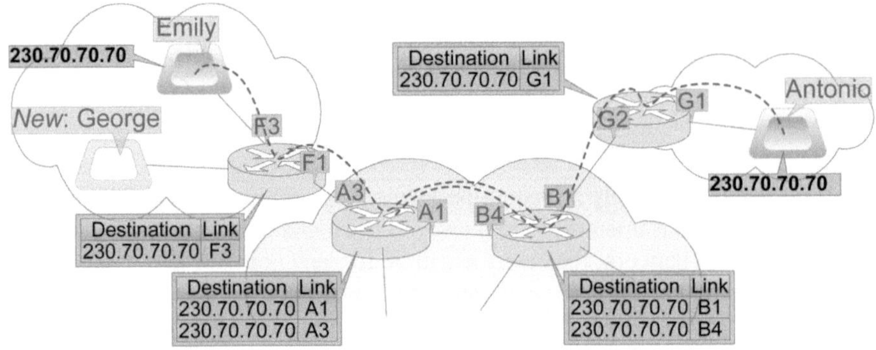

Fig. 7.12 Entry point discovery in a multicast network. Emily and Antonio advertize their membership to multicast group 230.70.70.70

including the individual address for replies. Based on its routing table, F forwards the query via link F3. Then, Emily responds back to George's individual address (Fig. 7.13a). In that way, the new node discovers its entry point and, at the same time, joins the multicast group.

Multicast is a networking protocol designed to reach every group member via a single query. For instance, when George issues a *join*, the query not only reaches Emily, but also propagates through routers F and A (Fig. 7.13a). Thus, when the discovery process relies on multicast, there is a chance to incur substantial traffic onto the IP network. One way to contain the group query propagation is by setting

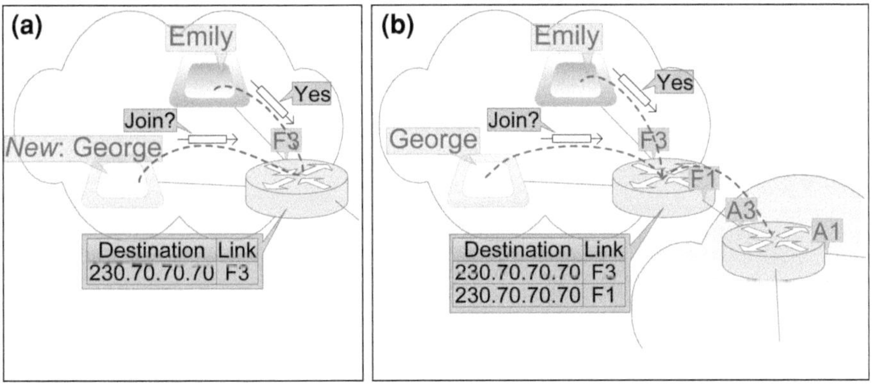

Fig. 7.13 George sends join requests to multiple members or a multicast group

a maximum time-to-live (TTL) parameter. For instance, a *join* from George set with TTL = 2 would only reach Emily, expiring as soon as it reaches router A. Obviously, if Emily was not sitting on F3, the query would fail. Therefore, George would have no other options than re-issue the *join* with a higher TTL value.

The latest version of the IP protocol, IPv6, supports yet another network mechanism known as *anycast* [16–18]. With anycast, the router can choose the closest member by itself, which helps optimizing bandwidth and response time. In fact, in most cases, the node *join* requires just a single entry point whilst multicast would reach all the terminals within the TTL horizon.

Another benefit of anycasting is that it considerably reduces the response time as the first router provides a response with no need to propagate the query through the groups. However, anycasting does not have any reporting mechanism; so when a discovery request is lost, the whole *join* fails and has to be restarted upon timing up.

Thus, while multicasting generates substantial traffic, anycasting may increase failure rates. A way to deal with these situations is use *multicast sub-groups*. An example is depicted in Fig. 7.14 whereby Emily has setup subgroup 230.70.70.71 while Antonio operates on 230.70.70.70. In order to join, George will have to multicast (or anycast) a request on one of the two multicast addresses. Thus his messages will traverse a smaller portion of the network.

7.8 Content is an Asset at the Edges

The concept of "virtual network" is a very powerful one and has indeed practical applicability. Virtual networks allow us to "reinvent" the network without having to modify the physical transport medium. In this chapter, we have explored one of the possible dimensions of virtual networks, that is, "content-awareness." We experience the power of content-aware networks any time we connect to a P2P

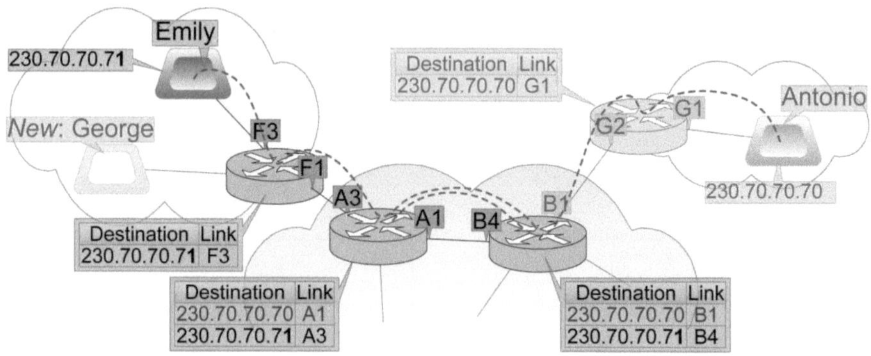

Fig. 7.14 Emily and Antonio join different multicast sub-groups in the same virtual network

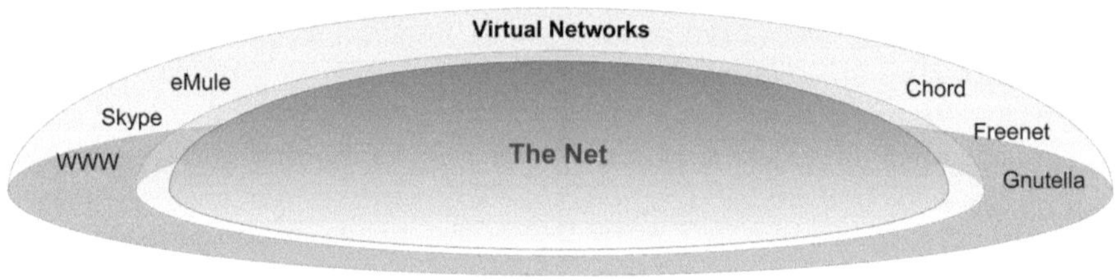

Fig. 7.15 Virtual networks rely on the Net to implement functionalities in the applications domain

systems. File-sharing applications eMule and Skype are just a tiny representation of content-aware networks, that is, ones that have redesigned the world of communications around what is most precious in today's digital ecosystem: *the content.*

The content is an asset that is produced and consumed at the edges of the Net (Fig. 7.15). Today, the Net has been relegated to a dumb bit-pipe; it moves raw data whose ultimate destiny is unknown. By contrast, P2P networks are specifically designed to boost up content consumption.

Thus far, we have seen how a virtual P2P network is born; how the ultimate beneficiaries, the peers, can join in. However, we have not yet discussed how the content is published on a virtual network, which will be the subject of the next chapter.

References

1. Miller M (2001) Discovering P2P. Wiley, New York
2. Oram A (ed) (2001) Peer-to-peer: harnessing the power of disruptive technologies. O'Reilly Media, USA
3. Vu QH et al (2009) Peer-to-peer computing principles and applications. Springer, Berlin

4. Buford J et al (2009) P2P networking and applications. Morgan Kaufmann, Burlington
5. Lua EK et al (2005) A survey and comparison of peer-to-peer overlay network schemes. IEEE Commun Surv Tutor 7:72–93
6. Knoll M et al (2008) Bootstrapping in peer-to-peer systems. In: IEEE international conference on parallel and distributed systems, pp 271–278
7. Dickey C, Grothoff C (2008) Bootstrapping of peer-to-peer networks. In: International symposium on applications and the internet, pp 205–208
8. Feder PM, Narvaez P (2009) Methods for initial bootstrap of user terminals in network. http://www.freepatentsonline.com/y2009/0292909.html
9. Bonfiglio D et al (2009) Detailed analysis of skype traffic. IEEE Trans Multimed 11:117–127
10. Eberspächer J, Schollmeier R (2005) First and second generation of peer-to-peer systems. Peer-to-peer systems and applications. Springer, Berlin, pp 35–56
11. Saroiu S et al (2003) Measuring and analyzing the characteristics of Napster and Gnutella hosts. Multimed Syst 9:170–184
12. Karbhari P, Ammar M, Dhamdhere A, Raj H, Riley G, Zegura E (2004) Bootstrapping in gnutella: a measurement study. In: Barakat C, Pratt I (eds) Passive and active network measurement, vol 3015. Springer, Berlin, pp 22–32. http://dx.doi.org/10.1007/978-3-540-24668-8_3. doi:10.1007/978-3-540-24668-8_3
13. Meshkova E et al (2008) A survey on resource discovery mechanisms peer-to-peer and service discovery frameworks. Comput Netw 52:2097–2128
14. Androutsellis-Theotokis S, Spinellis D (2004) A survey of peer-to-peer content distribution technologies. ACM Comput Surv 36:335–371
15. Clarke I et al (2001) Freenet: a distributed anonymous information storage and retrieval system. In: Federrath H (ed) Designing privacy enhancing technologies. Springer, Berlin, pp 46–66
16. Cramer C et al (2004) Bootstrapping locality-aware P2P networks. IEEE Int Conf Netw 1:357–361
17. Dinger J, Waldhorst O (2009) Decentralized bootstrapping of P2P systems: a practical view. In: International networking conference, pp 703–715
18. Yeo CK et al (2004) A survey of application level multicast techniques. Comput Commun 27:1547–1568

Chapter 8
Distribution–Efficient Networks

Abstract Publishing resources on a virtual network is a way to realize efficient data-distribution mechanisms. To this extent, each node needs to discover the other nodes, create neighborhoods and advertize its own resources. This chapter presents different techniques for making resources "discoverable," considering two approaches dubbed as *unstructured networks* and *structured networks*. We discuss properties of different protocols in terms of signaling overheads and distribution efficiency.

All truths are easy to understand once they are discovered; the point is to discover them
Galileo Galilei, Philosopher, Astronomer and Mathematician

8.1 Publishing Goes Beyond Bootstrapping

Identifying an entry point of a P2P Network is only the beginning of the fusion with other peers. After this initial step, the new peer has to improve its own visibility to the remaining network. This happens during the joining procedure which makes sure that peers and resources are published into the virtual network, becoming visible to the other peers.

Gaining visibility is a threefold objective of the "joining & publishing" phase. First, the new peer needs to gain fast and reliable access to other online nodes. Then, the online nodes must be able to discover the new peer. Finally,

A. Liotta and G. Exarchakos, *Networks for Pervasive Services*, Lecture Notes in Electrical Engineering, 92, DOI: 10.1007/978-94-007-1473-1_8,

any new resources introduced with the new peer must become discoverable too. In this chapter, we give an overview of various "join & publish" mechanisms, showing how different solutions impact the network in different ways. We shall see distinctive techniques in which the peers augment their neighbor lists, announce their presence with the network and publish their resources. Publishing resources on a virtual network is a way to realize efficient data-distribution mechanisms.

8.2 The Two Flavors of Virtual Networking

Different joining mechanisms impact the network structure in different ways. During the joining phase, a node populates the neighbor list, attaching more virtual links to other online nodes. Symmetrically, the new node attracts links throughout its lifetime. As links are added, removed or rewired, networks evolve and change properties, affecting the efficiency of *publish* and *discovery*. While the individual nodes populate and update their neighbor lists, the network changes form.

Any modification to a neighbor list is executed in one of the following two ways: *random* or *meticulous* [1–4]. While the former helps a peer to link to a random subset of online nodes, the latter creates links to very specific targets. Networks that grow randomly lack a deterministic structure and are therefore named *unstructured* networks. On the other hand, we distinguish the *structured* networks which evolve in a more controlled fashion.

The location of the actual resources within the network drives the organization of nodes and links, i.e., the *network topology*. If a resource exists in one or more replicas hosted by random nodes, a node does not have enough knowledge to build a guaranteed successful neighbor list. On the other hand, if the resources are located on deterministically chosen peer IDs, then the neighbor lists can direct definite messages to hosts.

The number of virtual links per node (i.e., *node degree*) plays an important role regarding the characterization of a network topology. For practical reasons, the size of a neighbor list (i.e. the *outgoing degree*) in structured or unstructured topologies is usually upper-bounded. Apart from the outgoing links, each node is reached by a number of *virtual links* created by the other nodes (i.e., the *incoming degree*). Incoming and outgoing links are two types of links that differ in roles. Nodes with a high outgoing degree may reach other nodes and resources fast and reliably. That is, a neighbor list pointing to several regions of a network brings the node closer to resources and reduces the probability of disconnection. On the other hand, a high incoming degree makes sure that a node is directly accessible from other online nodes. Thus, resources hosted in that node are also easily discoverable from multiple regions of the network.

8.3 Creating Unstructured Neighborhoods

Unstructured networks, in principle, consist of nodes that host and index their own local resources in addition to some other *remote* resources. Upon joining the network, new nodes try to collect a set of random online neighbors. Nodes do not delegate the *hosting* or *indexing* of their resources to other hosts. Therefore, a new node may only blindly select some neighbors or, in turn, be included in other neighbor lists.

Assuming that, after bootstrapping, a node has only found one neighbor, it will be necessary to trigger a *neighbor discovery* process to compile a meaningful *neighbor list*. If the nodes are building up a *flat unstructured* network having no prior knowledge, the new-coming nodes have no reason to give preference to any specific network element or to preclude certain IDs from their neighbor lists.

There are various ways in which nodes can build up their neighbor list. Let us take a closer look at the *Ping–Pong* method used in *Gnutella* [5]. Any new node (a *ping node*) sends out a *ping* message. On the other end, any node that receives the *ping* (a *pong node*) replies with a *pong* message to announce its presence. To avoid infinite loops and network flooding, *ping* messages expire after travelling through a given number of hops. To this purpose, the *ping* originator sets this threshold onto the Hops-to-Live (HTL) header field of the message. As the message travels from hop to hop, the hosting nodes progressively decrement the HTL value or drop the packet if HTL has reached zero. *Ping* messages are also dropped if the hosting node detects that they are repeated replicas that have previously been processed (the IDs of the *ping* messages are cached in the nodes for this purpose). Figure 8.1 illustrates the *ping* propagation process starting from node T. For simplicity, the *pong* replies are hidden in this illustration.

Besides relaying *ping* messages, every node replies with a *pong*. *Pong* messages travel backwards through the propagation path previously created by their *ping* counterparts. This process is illustrated through an example in Fig. 8.2, where some *ping* messages are hidden for simplicity (refer to Fig. 8.1). *Pong* messages provide information about the way the *ping node* can open a connection and transfer data. For instance, the IP address and port number to which the *pong node* can accept connections are the two main fields carried by a *pong*. Besides connection details, some nodes may inject extra information about their status (underloaded or overloaded), bandwidth, number and type of shared resources, etc.

Without the *pong* mechanism, the *ping node* would remain agnostic of any online nodes. Though *pong* messages are destined to the *ping node*, intermediate nodes of their path may benefit from the information they carry. For instance, intermediate nodes may replace broken virtual links in their neighbor lists.

To contain flooding during the *pong* process, *pong* messages are also set with an HTL threshold. At their inception, *pong* messages are set with an HTL equal to the

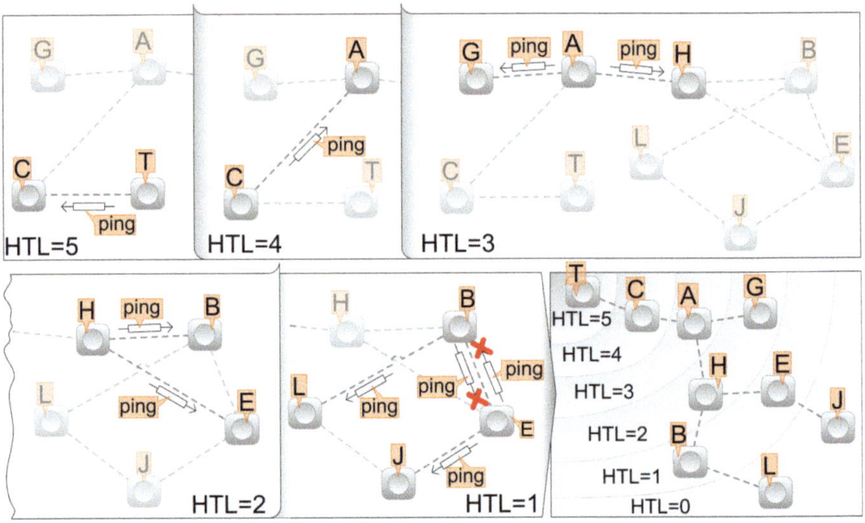

Fig. 8.1 *Ping* message propagation at maximum five hops. New node T starts a *ping* towards its entry point C. HTL decreases from five to four and node C relays it to neighbor A. Node T is excluded from this relaying step as the *ping* originator. Nodes A, H, B and E repeat the process. Node G receives the message with HTL bigger than zero. However, due to the lack of extra neighbors, it stops any further propagation. Nodes B and E exchange and mutually reject *ping* messages as already processed. HTL gets exhausted on nodes J and L which are also the leaf nodes of the propagation tree. As *ping* messages fork to more and more nodes at every iteration, and since loops are pruned as soon as they are detected, the propagation paths expand into tree-like structures

hop-distance from their corresponding *ping nodes*. The example of Fig. 8.3 assumes that the neighbor list contains at least eight entries (i.e., eight virtual links). Thus, node T can accommodate all *pong originator* IDs. There is also a mechanism to handle the situation in which a *ping node* receives more *pongs* than it can accommodate. Possible link-replacement schemes are "First-In-First-Out" and "Least-Recently-Used," though other strategies are also possible.

The *Ping–Pong* mechanism is able to detect any active node that is located within the HTL horizon of the *ping originator*. The main difficulty of *Ping–Pong* is choosing an "appropriate" value for HTL. Ideally, a node would like to discover a number of neighbors which is of the same magnitude of its *neighbor list* size. Discovering a number of neighbors that is greater than the neighbor list would simply waste resources (more *ping* and *pong* messages than necessary are incurred). On the other hand, if a node can only discover a very small number of neighbors, this will limit the level of parallelism of the virtual network.

The discovery horizon of a node is delimited by HTL. However, the relation between HTL horizon and the number of discovered nodes is not obvious. Clearly, the nodes that sit beyond the horizon cannot be reached through the *ping process* (i.e., *pings* are dropped at the horizon boundary). However, we cannot ensure that

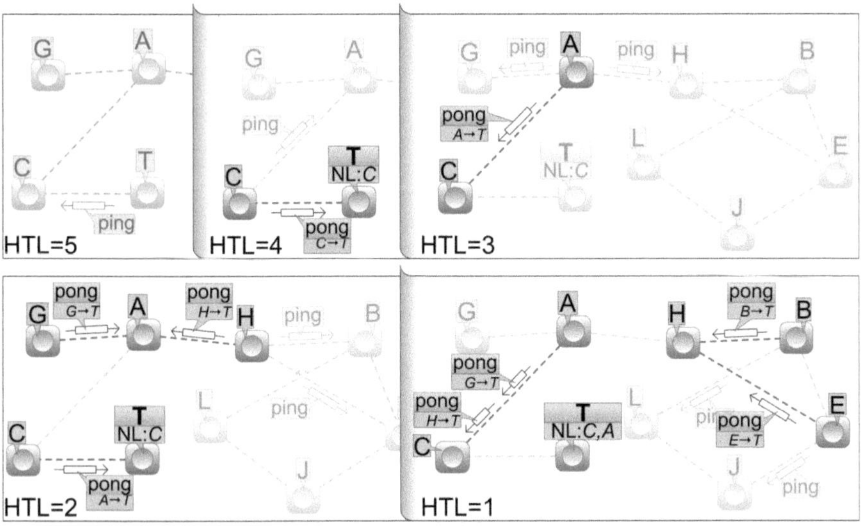

Fig. 8.2 The *pong* delivery process that sends a response back to the *ping* originator. This takes place in parallel to the *ping* propagation process of Fig. 8.1 (not shown here for simplicity). Upon receiving a *ping* message, node C replies with a *pong* (while the original *ping* has got HTL = 4). The originator (node T) enlists C into its neighbor list. On the next timeslot (HTL = 3), node A replies with a *pong* which travels backwards through the *ping path* (T → C → A). Similarly, *pong* messages from nodes G and H travel to *ping node* C via intermediate nodes A and C. Note that only some of the *pong* messages manage to reach their destination T before the *pings* have expired

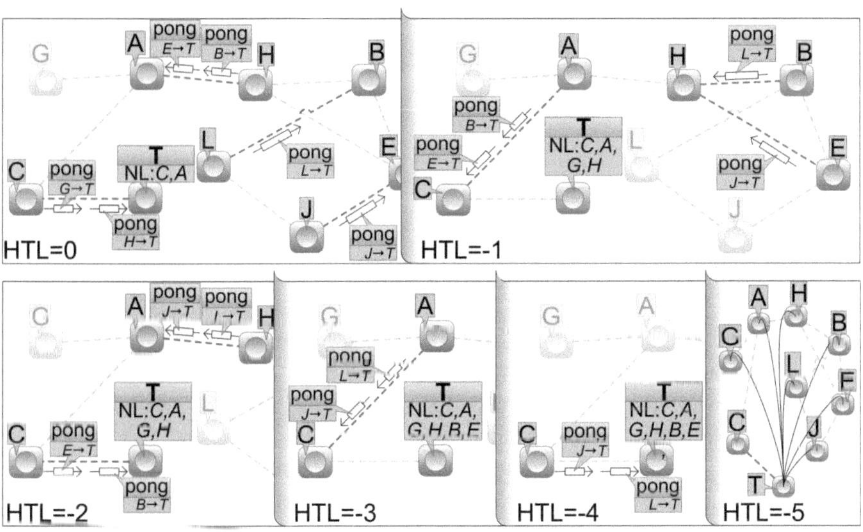

Fig. 8.3 *Pong* messages keep travelling even after the *ping* propagation has stopped. The last *pong* messages may reach the *ping* originator T even after 2*HTL steps from the first *ping* transmission

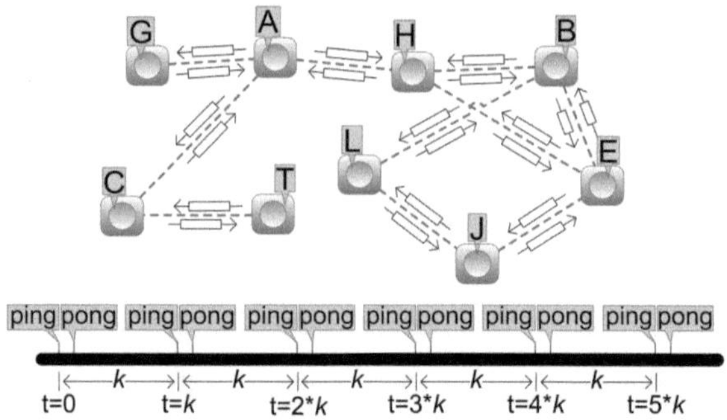

Fig. 8.4 *Cached Ping–Pong* iterations. All nodes send *pings* having HTL = 1 in order to ensure that these are not further relayed beyond the 1-hop neighbors. *Pongs* also have HTL = 1. *Pongs* carry the list of node IDs cached in the *pong node*. Upon receiving a *pong*, a node incorporates the new node IDs into its own cache

all nodes within the horizon are discovered. Also, the node *failure rate* has a negative impact on the discovery rate.

A variation of *Ping–Pong* is *Cached Ping–Pong* [6], a proactive approach that strives to reduce the number of unnecessary *pongs* and aims for faster node detection. All nodes send *pings* having HTL = 1 in order to ensure that these are not further relayed beyond the 1-hop neighbors. *Pongs* also have HTL = 1. *Pongs* carry the list of node IDs cached in the *pong node*. Upon receiving a *pong*, a node incorporates the new node IDs into its own cache. In this way, iteration after iteration, the catching area of each *pong* increases and all neighbors are gradually discovered. The periodic 1-hop *ping* is also an effective way to promptly discover link or node failures and propagate this knowledge in the subsequent iterations. An example is depicted in Fig. 8.4, whereas Table 8.1 shows the progression of knowledge propagation throughout the network.

8.4 Making Yourself Known in Unstructured Neighborhoods

Ping–Pong and *Cached Ping–Pong* are designed to help nodes collect online IDs. Once the *ping node* has created a bi-directional link to the *pong node*, requests can be relayed from and to both sides. However, this approach is not effective in networks that merely support unidirectional links. In this case, the joining node must "push" its advert more explicitly. The *announcement path* method [7] depicted in Fig. 8.5 uses *hello* messages instead of *pings* and does not require *pongs*.

The announcement path mechanism does not comprise a complete solution to the node joining procedure. It has to be coupled with another mechanism that populates a node's neighbor list. Otherwise, the new node may still achieve a low refresh rate of its neighbor list by participating in announcement paths. As soon as a node joins the network, the only known neighbor is its entry point. Through that

Table 8.1 *Cached Ping–Pong* messaging relating to Fig. 8.4

	IDs carried on pongs at certain timeslots					
From	t=1	*k*+1	2*k*+1	3*k*+1	4*k*+1	5*k*+1
A	A	A,**C**,**G**,**H**	A,**B**,C,**E**,G,H,**T**	A,B,C,E,G,H,**J**,**L**,T	*-all-*	*-all-*
B	B	B,**E**,**H**,**L**	**A**,B,E,H,**J**,L	A,B,**C**,E,**G**,H,J,L	*-all-*	*-all-*
C	C	**A**,C,**T**	A,C,**G**,**H**,T	A,**B**,C,**E**,G,H,T	A,B,C,E,G,H,**J**,**L**,T	*-all-*
E	E	**B**,E,**H**,**J**	**A**,B,E,H,J,**L**	A,B,**C**,E,**G**,H,J,L	A,B,C,E,G,H,J,L,**T**	*-all-*
G	G	**A**,G	A,**C**,G,**H**	A,**B**,C,**E**,G,H,**T**	A,B,C,E,G,H,**J**,**L**,T	*-all-*
H	H	**A**,**B**,**E**,H	A,B,**C**,E,**G**,H,**J**,**L**	A,B,C,E,G,H,J,L,**T**	A,B,C,E,G,H,**J**,**L**,T	*-all-*
J	J	**E**,J,**L**	**B**,E,**H**,J,L	**A**,B,E,H,J,L	A,B,**C**,E,**G**,H,J,L	*-all-*
L	L	**B**,**J**,L	B,**E**,**H**,J,L	**A**,B,E,H,J,L	A,B,**C**,E,**G**,H,J,L	*-all-*
T	T	**C**,T	**A**,C,T	A,C,**G**,**H**,T	A,**B**,C,**E**,G,H,T	*-all-*

Pongs are sent one timeslot after their corresponding *pings*, e.g., at timeslot $t = 0$, all nodes send a *ping*; at $t = 1$, they receive a *pong*. Assuming no prior exchange of *pongs*, at $t = 1$, nodes can only *pong* their own IDs; this is an important step as nodes can detect broken links. At the following exchanges, *pong* messages carry an increasing load of neighbor's neighbors. Our sample network has a 5-hop diameter, and thus all nodes are discovered in five iterations. For example, the colored cells of the table illustrate the path of ID = "J" hopping from node J to the new node T via nodes E, H, A and C

entry point, the node may receive announcements of other newly joined nodes and link to them. Thus, in that case, the joining procedure highly depends on the join rate of new nodes.

8.5 Unstructured Resource Publishing

Without any control over the ownership of resources, unstructured networks cannot relocate newly joined content closer to requestors. On the other hand, owners usually do not outsource resources, but shorten the distance from the requestors by spreading "adverts" on the network. Resource indexing on a single server (such as in *Napster* [8]) is a form of advertisement on a central place, namely, the *blackboard*. Every new node sends an index of its resources to that board so that the requestors can easily extract useful content.

The placement of adverts into multiple hosts across the whole network can address the robustness and scalability limitations of a central server. The closer a resource advert is to a requestor, the less the latency and signaling are. Replication of the same advert in many places can satisfy even quite distant requestors. Figure 8.6 illustrates a simple example of how adverts may improve the discovery performance.

However, advert replication strategies involve a number of risks. Although the discovery performance increases with the number of adverts dispersed across the network, the process of updating multiple adverts would generate massive bursts

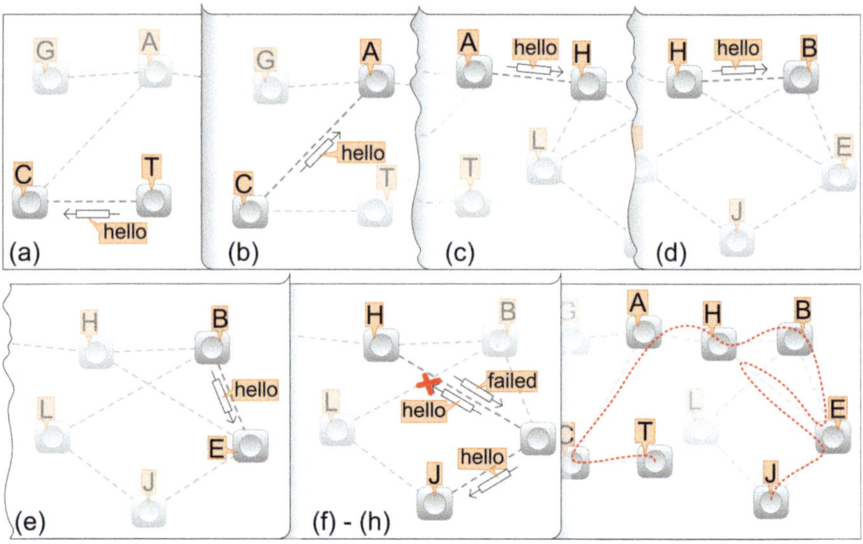

Fig. 8.5 *Announcement path* method. **a** Newly joined node T randomly selects one of his entry points (these have been discovered as part of the *joining* process) to announce itself with a *hello* message (in this case, there is only one choice, i.e., node C). **b** Node C adds T to its neighbor list, decrements the *hello*'s HTL and relays the *hello* to yet another randomly chosen neighbor (suppose HTL is still greater than zero at this point). **c–e** The same process continues at nodes A, H and B. **f–h** Node E happens to choose H, but now the process changes because H has already been visited. To avoid loops, H notifies its predecessor, node E, of this anomaly. Node E randomly picks another neighbor (J in this example) and resumes the path construction process. However, at J, the process stops when HTL reaches zero. Eventually, all the nodes included in the visited path have a direct virtual link to node T

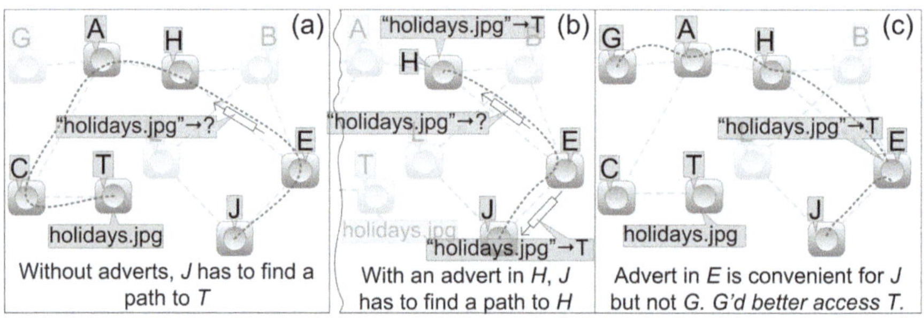

Fig. 8.6 Resource advertisements dispersed throughout the network may shorten the discovery paths. **a** Node J requests for "holidays.jpg," which is hosted in node T. Resource discovery induces a number of messages and long latencies. **b** An advert in node H relays the incoming requests directly to the host, reducing the signaling latency. **c** Advert locations play an important role in discovery performance

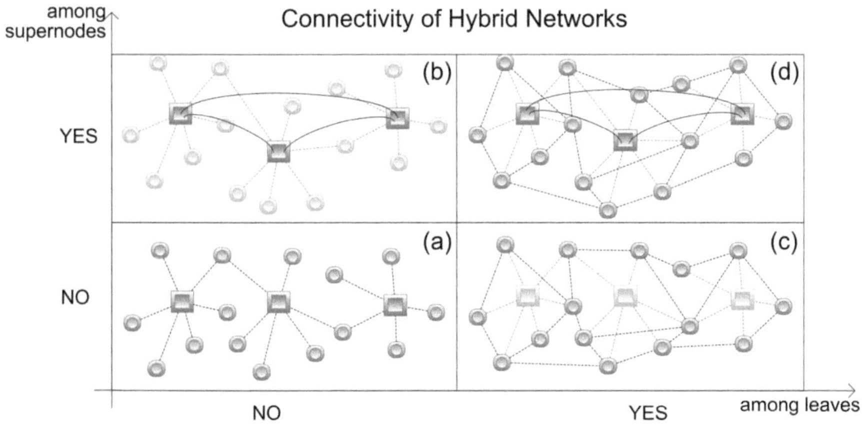

Fig. 8.7 The topology of Hybrid Networks changes based on the presence of links between supernodes and between leaves. **a** Supernode-to-supernode or leaf-to-leaf links are not allowed; the system is a set of Napster-like networks. Communication between supernodes is only indirectly possible via their common leaves. **b** Adding connections between supernodes allows indirect communication among all leaves via the supernode overlay. **c** Direct leaf-to-leaf connections make the system robust to supernode failures as mechanisms deployed in flat organizations could substitute the indexing services of supernode overlay. **d** If both types of connections are allowed, both robustness and efficient discovery issues are addressed

of traffic. To increase the adverts accuracy, both the resource and the advert hosts need to deploy frequent updates, causing serious scalability issues. Hence, the adverts should be as many as necessary and carefully placed in the network so that they reduce the overall discovery cost. Advert producers need to have some initial knowledge of their network size to optimize these parameters. However, in large-scale networks, this is nearly impossible.

To facilitate these decisions, both resource providers and requestors may use *advert boards*, that is, "*rendezvous*" points. Providers may choose one or more of these points to advertise resources. Unlike a *Napster* server, the "*rendezvous*" nodes (i.e., supernodes) index resources only hosted by a subset of nodes (the leaves). Besides normal activity and advert indexing, these supernodes may even take over extra roles such as bootstrapping. Upon receiving a request, they check their resource index to identify leaf hosts that could satisfy the request, returning a set of node IP addresses.

Peer-to-peer networks that introduce multiple centralization entities having only a partial view of the global knowledge are known as *Hybrid Networks* [9–11]. The two factors that affect the network topology are the connectivity between leaves and the connectivity between supernodes, as shown in Fig. 8.7.

Resource providers only need to know a small subset of supernodes which are also known to requestors. Though supernodes introduce a certain level of centralization, the network does not suffer from a single point of failure. There is a

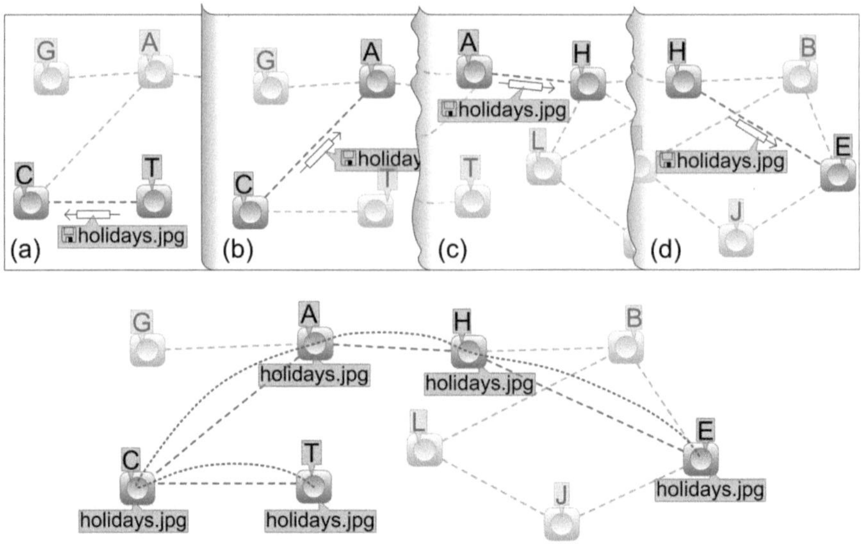

Fig. 8.8 Replication path of newly joined resources. New node T starts an *insert* request with HTL = 4. **a** The path starts from the joining node via its entry point, C. **b** Node C has no other neighbor; thus, the *insert* request is relayed to node A. **c** Considering the two neighbors of A, G and H, the letter "H" is alphabetically closer to the key "holidays.jpg." The file is replicated to node H. **d** The file "holidays.jpg" is also copied to node E and not B for similar reasons

plethora of such nodes with redundant information as leaves publish their adverts to more than one point. Symmetrically, leaves do not rely on a single supernode as they maintain a list of them to access either in parallel or sequentially. Hybrid networks address the issue of cost–effective and fast resource discovery by enhancing the role of a subset of nodes with indexing capabilities. However, in large-scale networks of intermittent nodes, it is difficult to ensure the optimal number of supernodes without an expensive monitoring system.

Instead of advertisements, *Freenet* goes one step further by replicating newly joined resources along a single path of nodes with IDs similar to resource keys. An *insert* (replication) request travels from the new node on the path of hosts picked to have the closest IDs to the replicated resource key. Practically, an insertion message propagates in the same way as the *Freenet* announcement path analyzed before. If a node has already processed the same request, it notifies its predecessor in the path to pick the next closest neighbor. Figure 8.8 gives a step-by-step example of this mechanisms.

The replication of the file itself, instead of an advert, decouples resource availability from the presence and robustness of its original provider. The replication path construction mechanism prioritizes neighbors based on a node ID to resource key closeness scheme. It tries to place resources not to random hosts, but to those that the requestors can guess. Though neighbor lists remain random, the resource location starts getting a more structured form.

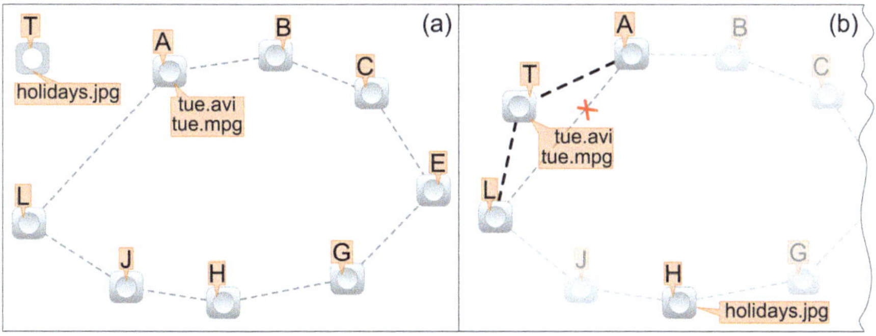

Fig. 8.9 Sample "structured" network built based on three rules: (1) every new node creates one link to the alphabetically next online ID; (2) the first letter of a resource name is the same as its host ID; (3) if such an ID is not available, the alphabetically next online node becomes the resource's host. Accordingly, files named "tue.avi" and "tue.mpg" should be hosted by T and file "holidays.jpg" by H. **a** Before node T joins the network, files "tue.avi" and "tue.mpg" are mapped onto the next online ID, i.e., node A. **b** After node T joins (connecting to its successor A) and node L relinks to T, the three files can be placed in the appropriate nodes, H and T, respectively

8.6 Secure a Role in Structure Worlds

Unlike the "unstructured" networks, their "structured" counterparts deploy a distributed mechanism that maps resources to node IDs [12]. This mechanism is usually a hash function, which produces a key for each resource in the network. These keys are values picked, as uniformly as possible, from a predefined space. Each node is responsible for a portion of the keys space (subspace). Thus, a node manages all the resources whose keys are mapped onto its subspace.

After obtaining a unique ID from the key space (bootstrapping phase), choosing an ID that is not yet in use, a node has to take over the management of the key subspace related to his ID. Symmetrically, this new node has to delegate the management of its local resources to the nodes that hold the relevant resource keys. Figure 8.9 illustrates a simple resource-to-node mapping algorithm that can be used to form a *structured virtual network*. The deterministic way, which distributes the resources over the network, makes it possible to know the host ID of a requested resource even before the discovery mechanism is triggered. The "structure" of the network is such that a search message can be directed to the resource-hosting node in a finite number of steps.

In general, each new node uses a hash function to map one of its "physical" IDs (e.g., its IP address or public PKI key) onto a node ID pulled from the key space. For instance, the scenario of Fig. 8.9 adopts the letters of the Latin alphabet as key space (Fig. 8.10). If some node IDs are missing, the existing ones take charge of the "orphan" resources, as illustrated in Fig. 8.11a. As new nodes join in, the key space is reorganized to spread the load more evenly, as illustrated in Fig. 8.11b. Due to its circular lined structure, this very first example of a structured virtual network is also dubbed the "Alphabelt."

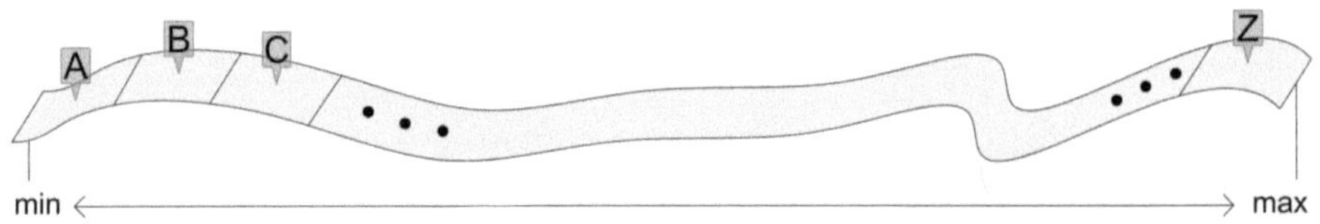

Fig. 8.10 Linear key space based on the Latin uppercase alphabet with 26 keys. There can only be a maximum of 26 nodes based on this key space. All resources are mapped to one of these 26 nodes based on the first letter of their name

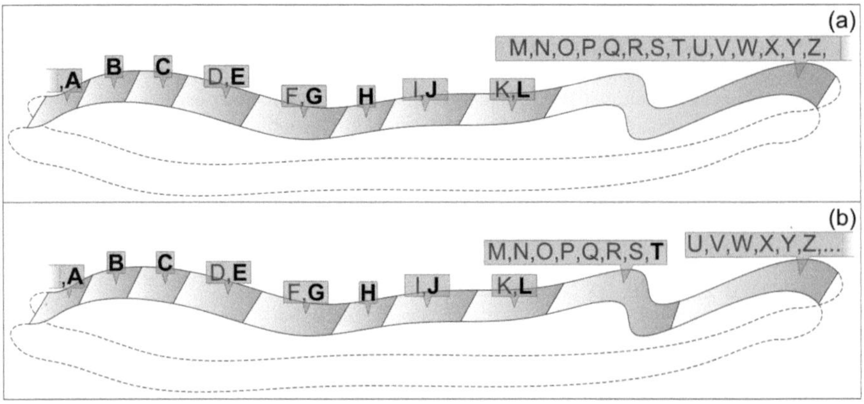

Fig. 8.11 The Latin alphabet key space is split into subspaces assigned to the nodes of Fig. 8.9. **a** The ID of a node (letters in *bold*) is the last key of the subspace it handles. As there is no node within the subspace [M, …, Z], the smallest online ID, i.e., node A, becomes the host of resources [M, …, Z]. **b** Node T joins in and the subspace [M, …, A] is further split into [M, …, T] and [U, …, A]

Once all key space slots are occupied by online nodes, no more nodes can join the network. To address this problem, structured networks commission much bigger key spaces than the number of nodes they anticipate. The key space should be big enough so that the probability that a node tries to acquire an occupied ID is negligible.

There is a variety of different key spaces in terms of size and type. The one presented above is a one-dimensional space, but extending it to multiple dimensions is a straightforward task [4].

8.7 Build Strict Formations

Once a node has generated its own ID, two more actions are required to complete the joining process. Based on its ID, the new node has to (1) find the appropriate location within the key space and (2) create virtual links with the nodes which

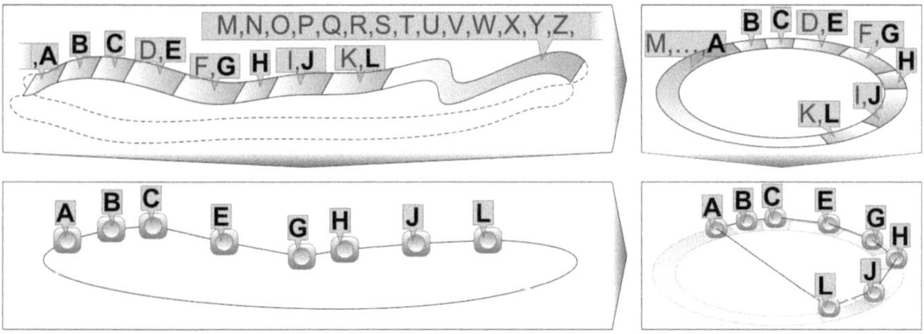

Fig. 8.12 This linear alphabet-based key space is mapped to a ring-connected set of nodes, each managing a subspace. Messages can only travel in a clockwise direction from one node to the next

manage neighboring subspaces. Practically, the first action consists of discovering the current host of the new ID and transferring the right portion of the subspace to the new node. At the end of this phase, the new node tries to fill in its neighbor list with more virtual links.

The exact procedures to execute these phases vary among different structured networks. In the case of the *Alphabelt network*, we sequentially map the nodes onto the linear key-space, forming a ring (Fig. 8.12). Eventually, every node has one neighbor, i.e., its successor.

Following a similar pattern from previous examples, at the bootstrapping phase, any new node needs to hash its address to produce a virtual ID and discover an entry point. If the entry point does not manage the new ID, it will function as a discovery proxy to locate another host who will handle the new virtual ID. Figure 8.13 provides analytic examples of the steps followed in order to join a ring-based structured network, that is:

- Step 1—Discovery of successor: the first node with ID bigger than or equal to the new ID
- Step 2—Connection to the successor
- Step 3—Retrieval of the predecessor
- Step 4—Notification to predecessor for re-linking to the new node.

The joining procedure described in the example above makes sure that there is exactly one path connecting any two nodes. Each node has just one entry in its neighbor list. These restrictions may seriously affect the robustness and discovery efficiency of the network. As shown in Fig. 8.14, a node failure would make it impossible to deliver messages between any of E's predecessors to any of E's successors. To address this problem, more virtual links per node are necessary. The next challenge is to decide how many extra links are required and how to establish them.

Starting from the extreme case of connecting all to all, the network results in a full mesh. If our online nodes are N in total, then all neighbor lists will have $N - 1$

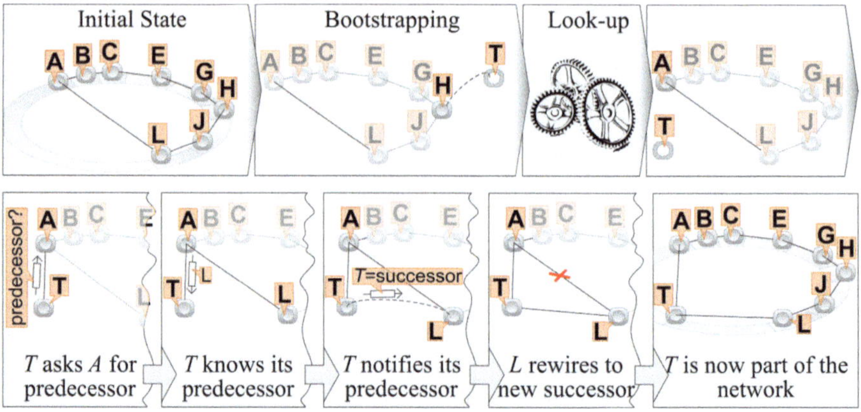

Fig. 8.13 Joining procedure of a new node in a ring-based structured network adopting a linear alphabet-based key space. The new node T bootstraps by hashing its IP address, generating virtual ID = "T" and then discovering node H as an entry point. Node H will seek for the current manager of "T," which is successor node A. Node T connects to A and asks for A's predecessor, which is node L. Then, node T notifies L that there is now a new successor. Node L breaks the L → A link and connects to new node T. Accordingly, the key space [M, …, A] is split in two, i.e., [M, …, T] and [U, …, Z, A]

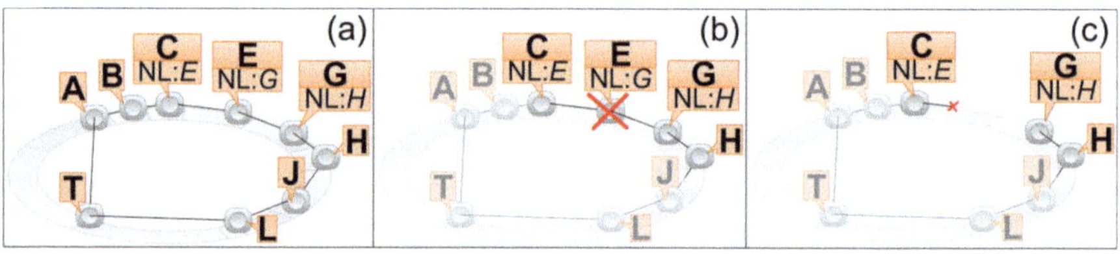

Fig. 8.14 A clockwise circular single-linked network suffers from node failures. Even one failure is enough to break the communication between two nodes. **a** Nodes have only one neighbor, i.e., the immediate higher ID. **b** Node E fails and the link E → G disappears. **c** Node C has a broken virtual link and any message originated from a node before E targeting a node after E cannot be delivered

entries. Requests can reach their destination node with just a single hop. However, the problems start when the nodes leave the network. For every failed node, $N - 1$ others need to detect their broken virtual links and update their neighbor lists. These networks need to deploy expensive maintenance mechanisms to keep their lists as accurate as possible. Due to this cost, full-mesh networks are not scalable, especially if they experience high *churn* rates (i.e., high frequency of node joins and departures) [13].

Structured networks try to find a compromise to this problem by applying a variety of approaches which strive to ensure finite-length paths between any two nodes. That is, nodes enlist neighbors in such a way that any request can

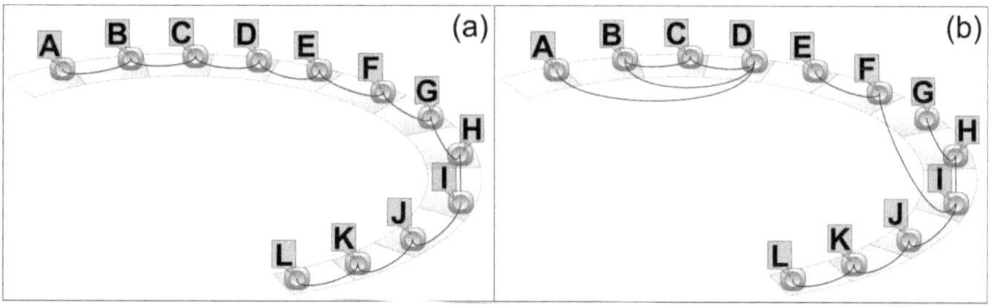

Fig. 8.15 Guarantees promised by a structured network may be broken if nodes have random neighbor lists. For clarity, only a part of the Alphabelt is visualized. **a** In the worst case scenario, a path from node A to L has to hop through 11 nodes in a densely populated key space. **b** Messages can get trapped in infinite loops. For instance, a request from node E looking for G in a clockwise Alphabelt with random neighborhoods would reach node I via F, overpassing target node G. The request would generate the infinite loop B → C→ D → B

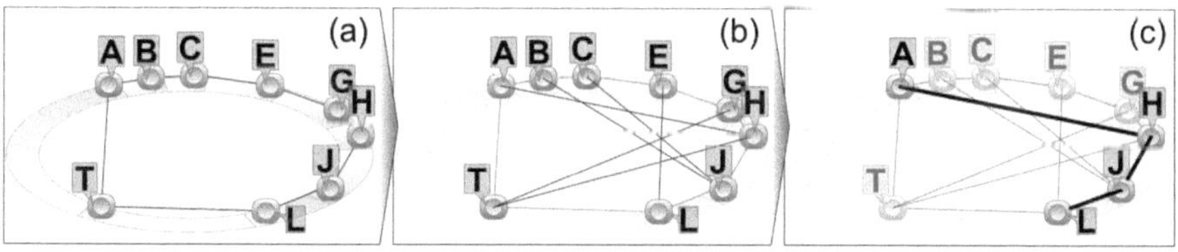

Fig. 8.16 Alphabelt with extra links between nodes in at least a quarter of key space distance. **a** A simple cyclic structured network. **b** Apart from links to immediate successors, nodes add one extra link to an ID at least 6.5 letters higher, e.g., A → H, B → J, C → J, etc. **c** This "boosted" connectivity helps to shorten the distance (hop-count) between two nodes, e.g., A → L is now four hops shorter than in case (**a**)

reach its destination via an upper-bounded number of hops. The general principle is to construct a path which guarantees a progressive approach to the destination node. We must guarantee that at each hop, the message gets closer to its destination. Figure 8.15 illustrates a case in which these guarantees are not provided.

Although there are a variety of network structures, in their very essence, they have a common feature: there is a path starting and ending at the same node that traverses all the others exactly once [14]. This kind of path is known as the *Hamiltonian cycle*. In the case of Fig. 8.15b, the connectivity among the nodes breaks the Hamiltonian cycle, violating the precondition for structured networks. This explains why the structured virtual network has to form a cyclic linked list. Adding more links, we can shorten the path, as messages can bypass a number of nodes and come closer to their target node. For instance, every node of our structured cyclic network could create a link to the node that holds an ID bigger by a quarter of key-space, as shown in Fig. 8.16.

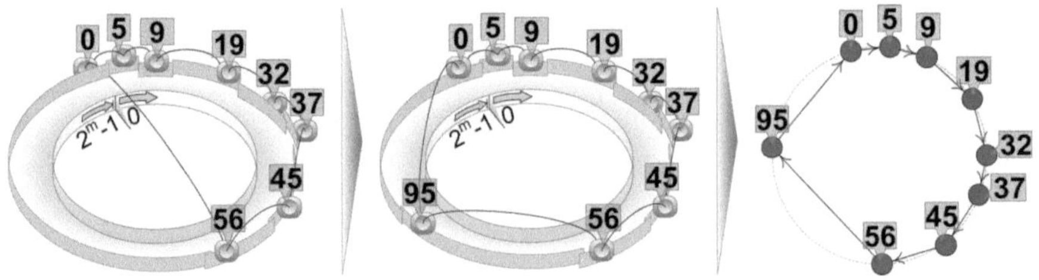

Fig. 8.17 A chord network of maximum $M = 128$ nodes, i.e., $m = 7$. It initially has $N = 8$ nodes, but after node 95 has joined ($N = 9$), we further split the key-space [0–127]. New node $n_{\text{new}} = 95$ uses its entry point $e_n = e_{95} = 37$ and the look-up function of the network to locate the host of key 95. This is the first node in the ring with ID bigger or equal to 95, i.e., node 0. Then, node 95 creates a virtual link with $n_{\text{successor}}(56) = 0$ and sends a message for exchanging keys and asking for the predecessor $n_{\text{predecessor}}(0)$. Thus, $n_{\text{new}} = 95$ triggers $n_{\text{predecessor}}(0) = 56$ to rewire to the new successor $n_{\text{successor}}(56) = 95$

8.8 Place Links and Resources into a Structured Ring

In addition to the *Alphabelt*, there are many more ways to realize structured networks. A widely used technique known as *Distributed hash table* (DHT) [1–4] creates an index of every resource and then keeps the index distributed across all online nodes. *SkipList*-based networks follow the approach to create multiple sorted and doubly-linked lists of different subsets of nodes grouped into levels [15–18]. Another possibility is to organize the nodes in tree-like structures [19] to maintain sorted resource indexes. We do not aim to provide a comprehensive analysis of all these types of networks. We wish to emphasize the potential of "structures" that place the resources into the right "bucket." It is interesting to see how distributed indexing works. For this purpose, we have chosen a well-known DHT-based system named "Chord" [2, 4, 20].

Like *Alphabelt*, *Chord* organizes nodes in a ring sorted clockwise on their ID. One could say that *Chord* is a more realistic and enhanced version of *Alphabelt*. Instead of Latin letters, *Chord* can accommodate *M* nodes with IDs spanning from *0* to $M - 1$. Node IDs are the output of a hash function on their corresponding IP addresses. If *m* is the number of bits of a node ID (expressed in binary form), then $M = 2^m$. Let us follow the example of Fig. 8.17 to illustrate *Chord*. Suppose we have *N* nodes ($1 \leq N \leq 2^m$). Thus, the lowest and highest possible IDs are *0* and $2^m{-}1$, respectively. A resource gets its key *k* via the same hash function, and hence from the same key space. The host of this resource is the node with $ID_{\text{host}} = k$ or the first node with $ID_{\text{host}} \geq k$.

The joining procedure of a new node (see Fig. 8.17) consists of two phases. The first is the discovery of a successor and predecessor. It is identical to the Alphabelt example with the exception of the look-up mechanism which varies. The second phase is the construction of the new node's *neighbor list*—in *Chord*, the neighbor list is actually named as the *finger table*. Nodes in Alphabelt had only one neighbor, the immediate successor; thus, this second phase was covered by the

i	$f_i(n_{new})=(n_{new}+2^i)mod(2^m)$, $0 \leq i \leq m-1$	$f_i(n_{new})$ Theoretical	Online
0	$(95+2^0)mod(2^7)=96mod(128)$	96	0
1	$(95+2^1)mod(2^7)=98mod(128)$	98	0
2	$(95+2^2)mod(128)=99mod(128)$	99	0
3	$(95+8)mod(128)=103mod(128)$	103	0
4	$111mod(128)$	111	0
5	$127mod(128)$	127	0
6	$159mod(128)$	31	32

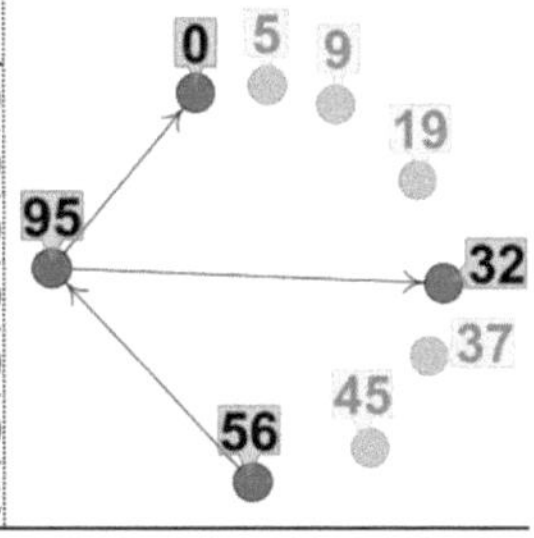

Fig. 8.18 *Chord* finger table of new node 95. Since the new node has ID = 95, the size of the key space cannot be less than 96 (0–95). Given that the key space must be $M = 2^m$ and $1 \leq N \leq 2^m$, then $m = 7$ and $M = 128$ (if $m = 6$, $2^m < 95$). Thus, the finger table must enlist seven *fingers*. Once that node 95 has bootstrapped and is aware of its successor, it calculates the theoretical fingers in its table. For each one of them, it triggers a look-up function to locate the node that hosts those theoretical keys. For example, the fifth position of the finger table is the theoretical: $f_5(95) = (95 + 2^5)\text{mod}(2^7) = 127\text{mod}(128) = 127$ hosted by node 0 as node 127 does not exist

first. However, nodes in *Chord* have to construct their finger with multiple virtual links.

The size of the nodes finger tables is equal to m. The ith ($0 \leq i \leq m - 1$) finger is an {ID, IP} pair of the $(n_{\text{new}} + 2^i)\text{mod}(2^m)$ node, i.e., $f_i(n_{\text{new}}) = (n_{\text{new}} + 2^i)$-mod$(2^m)$. If this function does not give an online node ID, the first successive online node is enlisted in the table instead. Every node calculates the theoretical fingers and then uses the look-up function to discover the online ones. The *Chord* discovery mechanism will always return a node with the requested ID or the first successive online one. Figure 8.18 illustrates a snapshot of the finger table of a new node.

A careful look at the finger table of Fig. 8.18 reveals the properties of the node 95 neighbor list. In general, the top entry of a *Chord* finger table is always the immediate successor; the last entry points to a node sitting $M/2$ keys away. Figure 8.18 illustrates the final status of all finger tables and the network structure with all virtual links connecting their node pairs. *Chord* is, in essence, a clockwise circular singly-linked list of nodes ordered on their ID—practically a directed graph. However, for graphical purposes, the *Chord* ring of Fig. 8.18 only shows undirected links—the actual directions appear in the finger tables given on the side.

Node 95 does not seem to be in the best position. There are $N - 1 = 8$ links from other nodes pointing at 95. Most likely, it is a hot destination node as it serves the key subspace [57, 95] (i.e., more than 30% of the whole key space). Moreover, node 0 appears in four finger tables, 5 in one, 9 in two, etc. By checking the size of their respective key space, we identify similarities with the number of their incoming links (*incoming degree*), as shown in Fig. 8.19. In general, we may assume that the bigger the key space a node handles, the more incoming links it will have.

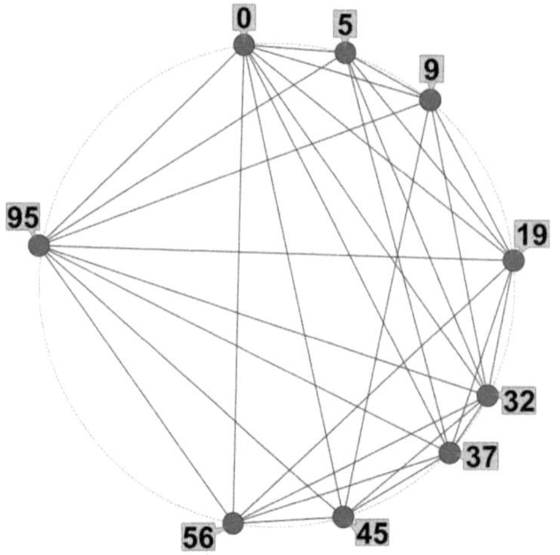

	0→*	5→*	9→*	19→*	32→*
i=0:	5	0: 9	0: 19	0: 32	0: 37
1:	5	1: 9	1: 19	1: 32	1: 37
2:	5	2: 9	2: 19	2: 32	2: 37
3:	9	3: 19	3: 19	3: 32	3: 45
4:	19	4: 32	4: 32	4: 37	4: 56
5:	32	5: 37	5: 45	5: 56	5: 95
6:	95	6: 95	6: 95	6: 95	6: 0

37→*	45→*	56→*	95→*
0: 45	0: 56	0: 95	0: 0
1: 45	1: 56	1: 95	1: 0
2: 45	2: 56	2: 95	2: 0
3: 45	3: 56	3: 95	3: 0
4: 56	4: 95	4: 95	4: 0
5: 95	5: 95	5: 95	5: 0
6: 0	6: 0	6: 0	6: 32

Fig. 8.19 Sample *Chord* network with the computed finger tables and links

ID	$ID_k - ID_{k-1}$	Degree
k=0	33	5
5	5	1
9	4	2
19	10	3
32	13	5
37	5	3
45	8	3
56	11	4
95	39	8

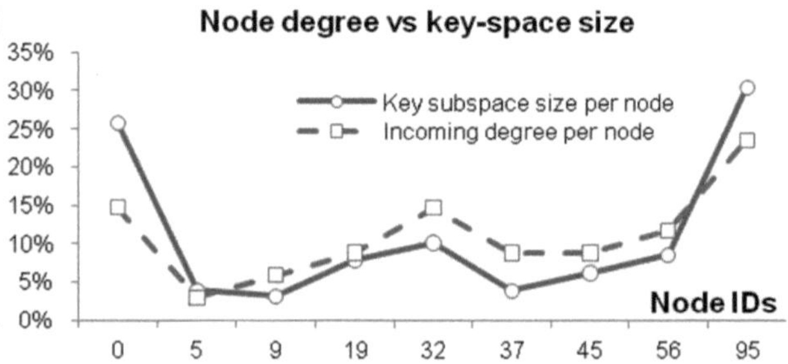

Fig. 8.20 Comparison of node degree against the key-space that is handled by the node. For instance, node 0 handles a key subspace 33 keys long and has 5 incoming links. The graph shows the percentage of incoming links and key subspace size per node over the total links and key space size, respectively

Figure 8.20 clarifies the effect that a hash function can have on the ring's *stability*, *balance* and *efficiency*. If the IDs are not uniformly distributed over the key space, some nodes might relatively host many more resources than others. The load of a node depends on the "popularity" and not necessarily on the "number" of resources it hosts. Even if a node handles a big portion of the global key space, the resources assigned to the node may be few or infamous.

In large *Chord* networks, discovery paths will be very long, affecting the *latency* and overall *performance* of the network. One way to address this issue is by realizing *hybrid Chord* networks [21, 22]. Extending the concept of "resources in the right bucket" to "resources and nodes in the right buckets," we end up with a *two-layered* structured network. Based on this idea, each node has a neighbor list pointing to the other leaf nodes of the ring and one link to a *supernode*. Each supernode indexes a subset of *Chord* nodes (leaves) and their key-subspace. Supernodes are also interconnected either via another *Chord* ring or mesh, or other types of topologies.

The global key-space is divided among supernodes which, instead of hosting resources, maintain the first and last node of the space they handle. Any new nodes, upon joining to the *Chord*, have to follow the same procedure as described above, but must also register with the supernode that manages their space. In this way, requests can traverse the whole *Chord* ring with only a few hops on the overlay of supernodes. The motivating factor is to reduce the search space by finding a quick way to limit the *Chord* look-up function within a supernode subspace. Thus, requestors can send requests to the supernode overlay. The supernode overlay submits each request to the first node of the appropriate subspace to execute the *Chord* look-up function on only a fraction of the global *Chord* ring.

Although the unstructured networks come short of discovery efficiency due to the randomness of their structure, they win in *flexibility* in case of fast network changes. Their joining process is much simpler and inexpensive as they do not trigger resources re-location upon the arrival of a new node. A limiting factor for structured networks is the key space that is statically set at the beginning of their lifetime.

8.9 Data-Awareness via Protocol-Agnosticism

New applications bring new types of heterogeneity in the network, trigger different user's behaviors, and introduce a plethora of new requirements. In reality, the connection between any two nodes is realized via physical (rather than virtual) links. These are tied to a wide range of network equipment installed worldwide which cannot be easily upgraded as soon as new application appears. This limitation is, in fact, addresses by virtual networks, whose functionality and protocols can be amended via software upgrades.

Thus far, the application developers have been trying to fill the mismatches between the emerging breed of software systems and the rigid networks that support their communication needs by re-implementing new communication protocols within the application itself (such as in P2P applications). These protocols are aware of the data they transmit, but are also specific to the applications they serve. Although these application-specific protocols are "data-aware" and help two nodes to understand each other, the underlying network delivers packets between any two points ignoring the kind of data being transferred. Being "data-agnostic," the underlying network cannot offer some of the essential network services that recent and future applications require. For instance, services like *discovery* or *filtering* of multiple resources based on some complex criteria or even smart bandwidth utilization based on the data transferred are not allowed.

All resource providers need to join and make the network aware, first, of their presence and, second, of their resources. It is only in this case that requestors can search resources without prior knowledge of their exact location. The techniques presented in this chapter are already applied to situations where content requestors completely ignore providers. There are protocols that satisfy the basic needs of any resource that has to become available: the resource is traceable and its host is discoverable.

Any new node aims at a robust connection with the network and publishing of local resources. These are aims that the node executes in three steps:

- expand the connectivity via populating its neighbor list,
- advertise, index or replicate its resources in the network, and
- let the network know about its presence.

There is a variety of mechanisms that implement these actions, all with their pros and cons. A network that tries to accommodate and optimally react on any data type and to satisfy any user requirement needs to stay agnostic of these mechanisms, i.e. it has to provide a generic enough framework able to assimilate any mechanism in place.

Every node has to populate its neighbor list with a number of other nodes that will make sure that their requests can travel deeper to other nodes within the network. There are two main ways to build these lists: (a) with random nodes or (b) with carefully selected ones. The first choice makes the network more robust to abrupt changes of nodes' stability as they can leave and join with very low impact to the network. However, building neighbor lists with purposely chosen nodes is a way to guarantee certain discovery performance. A new node has to contact, via its entry point, a number of other online nodes to fill up its list. To make sure that it is also accessible, it has to gossip its presence to random or deterministically chosen nodes, respectively. Unstructured node formations require resource publishing to ensure that one's resources are visible to requestors. Structured Networks can accurately define the location of the resources and deterministically drive requests to them; hence, there is no need for resource advertisement.

References

1. Vu QH et al (2009) Peer-to-peer computing: principles and applications. Springer, Heidelberg
2. Lua EK et al (2005) A survey and comparison of peer-to-peer overlay network schemes. IEEE Commun Surv Tutor 7:72–93
3. Buford J et al (2009) P2P networking and applications. Morgan Kaufmann, San Fransisco
4. Androutsellis-Theotokis S, Spinellis D (2004) A survey of peer-to-peer content distribution technologies. ACM Comput Surv 36:335–371
5. Ripeanu M et al (2002) Mapping the gnutella network. IEEE Internet Comput 6:50–57
6. Yang B et al (2004) Evaluating GUESS and non-forwarding peer-to-peer search. International Conference on Distributed Computing Systems, 209–218. doi:10.1109/ICDCS.2004.1281585
7. Clarke I et al (2001) Freenet: a distributed anonymous information storage and retrieval system. In: Federrath H (ed) Designing privacy enhancing technologies. Springer, Heidelberg, pp 46–66

8. Saroiu S et al (2003) Measuring and analyzing the characteristics of Napster and Gnutella hosts. Multimedia Syst 9:170–184
9. Loo BT et al (2005) The case for a hybrid P2P search infrastructure. Peer-to-peer systems III. Springer, Heidelberg, pp 141–150
10. Stutzbach D, Rejaie R (2005) Characterizing the two-tier Gnutella topology. ACM SIGMETRICS Performance Evaluation Review. ACM, pp 402–403
11. Exarchakos G, Antonopoulos N (2007) Resource sharing architecture for cooperative heterogeneous P2P overlays. J Netw Syst Manage 15:311–334
12. Datta A (2010) The gamut of bootstrapping mechanisms for structured overlay networks. Handbook of peer-to-peer networking. Springer, Heidelberg, pp 281–308
13. Li J et al (2005) Comparing the performance of distributed hash tables under churn. Peer-to-peer systems III. Springer, Heidelberg, pp 87–99
14. Qu C et al (2006) Cayley DHTs—a group-theoretic framework for analyzing dhts based on cayley graphs. Semantic web and peer-to-peer. Springer, Heidelberg, pp 89–105
15. Aspnes J, Shah G (2007) Skip graphs. ACM Trans Algorithms 3(37):1–25
16. Harvey NJA, Munro JI (2004) Deterministic SkipNet. Inform Process Lett 90:205–208
17. Abraham I et al (2006) Skip B-Trees. Principles of distributed systems. Springer, Heidelberg, pp 366–380
18. Naor M, Wieder U (2005) Know thy neighbor's neighbor: better routing for skip-graphs and small worlds. Peer-to-peer systems III. Springer, Heidelberg, pp 269–277
19. Aberer K et al (2003) P-Grid: a self-organizing structured P2P system. SIGMOD Rec 32:29–33
20. Stoica I et al (2003) Chord: a scalable peer-to-peer lookup protocol for Internet applications. IEEE/ACM Trans Netw 11:17–32
21. Garcés-Erice L et al (2003) Hierarchical peer to peer systems. Euro-Par 2003 parallel processing. Springer, Heidelberg, pp 1230–1239
22. Salter J, Antonopoulos N (2007) An optimized two-tier P2P architecture for contextualized keyword searches. Future Gener Comp Syst 23:241–251

Chapter 9
Discovering Virtual Resources

Abstract Virtual resources are considerably more numerous than the physical devices which host them. There are many more files than servers; more videos than people. Thus, the task of discovering relevant resources is certainly a daunting one. The *search engines* help to retrieve data from servers. Yet, discovering resources stored in a virtual environment requires "deep" searching techniques which must be able to explore not only the multitude of web servers, but also the ordinary computers and terminals. This chapter introduces different discovery techniques used in "structured" and "unstructured" P2P systems.

You affect the world by what you browse
Prof. Tim Berners-Lee, inventor of the World Wide Web

9.1 Four Ways to Reach a Resource

The Web contains a vast amount of information which is only useful if we can find it easily. Search engines such as *Google* or *Yahoo* do a great job of organizing the data stored in the myriad of the web server. Yet, how can we discover the plethora of files that are published and shared via virtual networks?

In Chap. 8, we discussed two ways to publish resources on a virtual network, the *structured* and *unstructured* P2P architectures. Similarly, we have two main classes of discovery algorithms, *structured* and *unstructured*, with a wealth of variations. Here, we explore a range of discovery mechanisms that fall under four different categories: *blind* discovery; *informed* discovery; *loosely structured* discovery; and *deterministic* discovery [1–5].

A. Liotta and G. Exarchakos, *Networks for Pervasive Services*,
Lecture Notes in Electrical Engineering, 92, DOI: 10.1007/978-94-007-1473-1_9,

9.2 Assessment of Discovery Mechanisms

Each discovery mechanism has its strengths and weaknesses. Each protocol incurs signaling overheads. However, given the distributed nature of "deep" searching, it is hard to know how far into the virtual network we have to reach, how long it will take to get a response, or whether we will actually get any response at all.

Thus, the performance metrics of "discovery" are rather peculiar. The discovery *efficiency*, namely, the *success rate*, is the portion of requests for which the discovery process manages to return at least one response. The discovery *accuracy*, that is, the *recall*, is the fraction of resource replicas that is actually discovered. Thus, if a file is replicated into 1,000 different nodes and the discovery process finds 100 replicas, the recall is 10%.

Increasing efficiency and accuracy comes with a cost in terms of signaling overheads and response time. These can be measured through the number of *signaling messages* generated and relayed, and via the *hop count,* that is, the number of hops separating the requestor from the nearest resource instance.

9.3 Containing the Proliferation of Discovery Messages

The number of "deep" search messages can easily proliferate, loop and get out of control. In general, there are three simple ways to contain the discovery process known as HTL, TTL and caching. The first approach is to set an upper limit on the number of nodes that the messages can visit before being purged. For this purpose, the request originator sets the Hop-to-Live (HTL) message field with an integer value which is decremented by each visited node. It is hard to determine a suitable value for HTL, unless we know the size of the network. Thus, some systems start off with a tentative HTL value which is subsequently increased until the required *success rate* and *recall* are achieved.

For the same purpose, other search algorithms use the Time-to-Live (TTL) instead of HTL. The TTL represents the maximum time that a requesting node is willing to wait for a response. Each visited node compares the TTL value of the message with the current time and decrements TTL by the amount of delay incurred on the last hop before relaying the message. Obviously, if the resulting TTL is zero, the message is purged instead of being relayed.

In principle, TTL is more accurate that HTL as it accounts for the variability of delays across different links. However, the implementation of TTL is more complex as it requires that all nodes synchronize their clocks. Thus, in practice, the TTL is often used in the same way as the HTL, accounting for the number of visited hops instead of the actual travel time of the messages.

Neither HTL nor TTL can prevent the messages from looping. During the propagation of the request, some messages may reach the same node twice. This leads to a waste in bandwidth and CPU cycles, with overall performance degradation. To prevent loops, the nodes cache the header of each message on the first visit. In this way, any revisit is immediately detected and dropped. Clearly, not all messages can be cached forever. Thus, each cache entry has a limited lifetime.

9.4 Blind Discovery for Unstructured Networks

When the resources are stored over an unstructured network, the discovery process cannot rely on prior knowledge. The search messages hop from node to node, following different variations of what is, in essence, a "random" process. Depending on the number of neighbors involved in relaying the discovery messages at each step and on the particular ways that the requests are copied in between neighbors (e.g., in parallel or sequentially), we distinguish a variety of "blind" discovery mechanisms [3, 5]. Hereafter, we describe five among the most popular ones, including *Breadth-first search* (BFS), *Modified breadth-first search* (mBFS), *Depth-first search* (DFS), *Iterative deepening search* (IDS) and *Random walks* (RW).

The Breadth-first search (that is, *flooding*) [6] explores every node within the HTL horizon. A requestor sends the same request in parallel to every immediate neighbor. Upon receiving a request, each neighbor reduces the HTL by one and forwards replicas to all its own neighbors. Hence, parallel propagation paths start from the request originator and fork at every hop. The request originator and the latest predecessor of a path are excluded from propagation.

We have already seen this algorithm in action in Chap. 8 when we discussed the Ping–Pong mechanism. "Ping" messages travel on a BFS propagation tree looking for online nodes instead of resources. As every visited node replies with a "Pong" message, the *success rate* of the Ping–Pong mechanism is 100%. Therefore, the *recall* is given by the ratio between the number of nodes contained within the HTL horizon and the size of the network. BFS is exemplified in Fig. 9.1.

BFS is a search mechanism that guarantees the discovery of any resource instance located within the HTL horizon. Hence, the *success rate* for popular resources can be quite high, as the probability to find one instance within the requestor's vicinity is also high. On the contrary, BFS has poor performance on rare resources which may actually lie outside of the HTL horizon. For similar reasons, the *recall* of the BFS mechanism is better on popular than on rare resources. Another problem is that, due to its tree-like exploration, BFS messages visit (or even revisit multiple times) all the nodes within the HTL horizon. Thus, the number of messages grows considerably with HTL.

In an attempt to reduce the message costs of the BFS, the *modified Breadth-First Search* (mBFS) [7] protocol limits the forwarding scope choosing, at each iteration, a random subset of the immediate neighbors. The size of this subset (W) is predefined by the network designer and is kept constant across the network. mBFS aims at visiting most of the nodes within the HTL horizon while at the same time trying to reduce loops. In this way, the messaging is reduced in comparison to BFS, without having to compromise the success rate and recall. However, the average number of hops traversed before a resource is discovered may increase. Also, as HTL grows, mBFS tends to perform just like BFS since more requests get

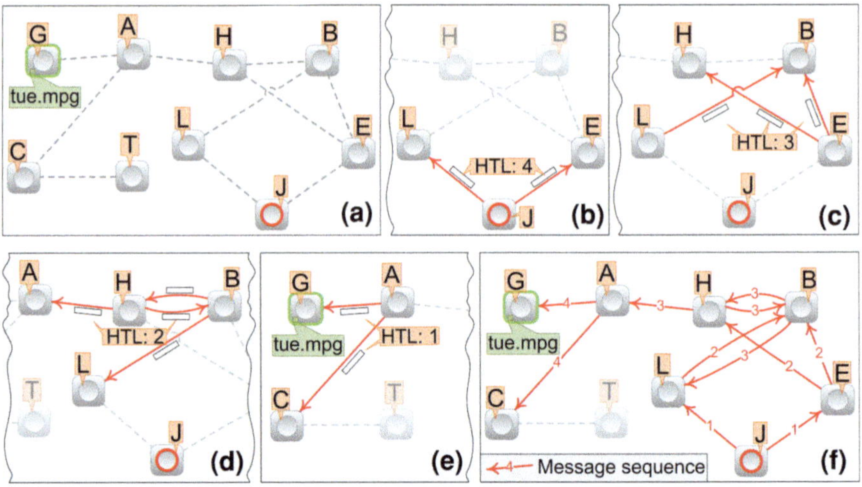

Fig. 9.1 Breadth-First Search (a.k.a. flooding). **a** Node J is looking for file "tue.mpg." **b** Initially, the hop-to-live is set to HTL = 4. **c** Nodes E and L decrement the HTL. **d** The process continues with a further HTL decrement. Two messages are wasted: node B forwards the request to H and L even though these have previously processed the request. This is because B is not aware of the link H → E and H is not aware of the link B → E. Even though node B receives two requests on the same timeslot (HTL = 3), the requests are practically buffered and processed on a first-come-first-serve basis. Assuming that the request from E is queued before the one from L, node B will prepare two replicas to forward to L and H. As soon as a request from L arrives, node B will reject it as already processed. **e** Out of B, H, L and A, only node A continues the process. Finally, the request reaches node G. **f** Complete message sequence

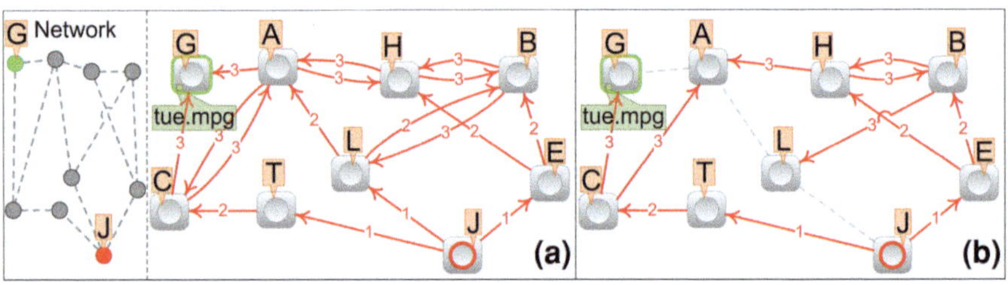

Fig. 9.2 Message propagation trees in **a** BFS and **b** mBFS. BFS generates many more messages than mBFS, which only selects a subset of the available neighbors during the flooding-based discovery process ($W = 2$ in this example)

into cycles. An example showing the different behavior of the mBFS and BFS is depicted in Fig. 9.2.

BFS and mBFS perform well in terms of *recall* and *hop count*, but incur heavy messaging. To reduce traffic, Depth-first search (DFS) [8] explores the network sequentially rather than in parallel, as exemplified in Fig. 9.3. The order in which the successor neighbors are selected may be random or, in other cases, it could respond to some kind of prioritization algorithm set at design time. Each

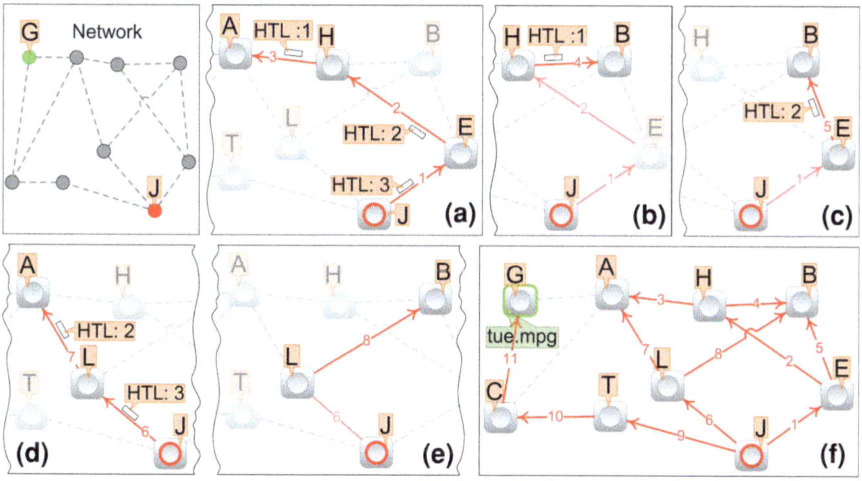

Fig. 9.3 The sequential network exploration with DFS and caching to contain the revisits. **a** The request travels over a single path until HTL expires. **b** Then, we backtrack on the last visited node (H in this case) and explore another alternative (node B). **c** After all descendents of H have been unsuccessfully explored, we further backtrack (node E), which leads again to negative outcomes. **d** The request backtracks up to the source (node J) and restarts in a new direction (node L and then A). **e** Node A has already processed the request and backtracks to L even if it was close to the resource provider G. Node L redirects the request to B. If caching were not activated, node A would explore B and G, or just G and would terminate. **f** The latest attempt hits a resource and the whole process stops

exploration path is HTL deep and the process stops as soon as a node is found to contain the requested resource.

In the worst-case scenario, that is, one in which the resource is not present within the given HTL horizon, DFS may end up exploring the same space as BFS. In general, however, the DFS *recall* tends to be lower than in BFS and the *hop count* is normally higher. In fact, there is a chance that DFS explores a whole subtree and discovers a resource which is located further away, missing the nearby resources located on a separate subtree. While DFS stops upon the first hit, BFS keeps expanding until HTL expires.

One way to combine the benefits of BFS with those of DFS is through the *Iterative deepening search* (IDS) [9] protocol, as exemplified in Fig. 9.4. The originator launches a sequence of BFS processes with increasing HTL values until a resource is found. After a request times out unsuccessfully (i.e., no resource is found within the HTL horizon), the requestor increases HTL and launches in this way a "deeper" BFS "wave." The main issue of this approach is that all the nodes visited on a certain wave are revisited during all the successive ones. To avoid useless messages and prevent response duplicates, resource providers reply only on the first wave that discovered them; any request from a subsequent wave is simply further relayed.

The tree-exploration algorithms described so far are strong in terms of *success rate* and *recall*, but incur high discovery costs which exponentially increase with

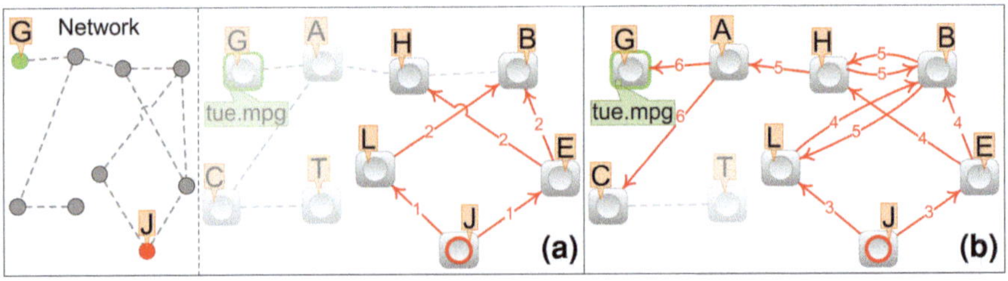

Fig. 9.4 Iterative deepening search on two subsequent waves. **a** First wave has HTL (i.e., depth) equal to two. The resource stored in node G cannot yet be discovered. **b** The second wave is increased to a depth of four and the resource "tue.mpg" is discovered

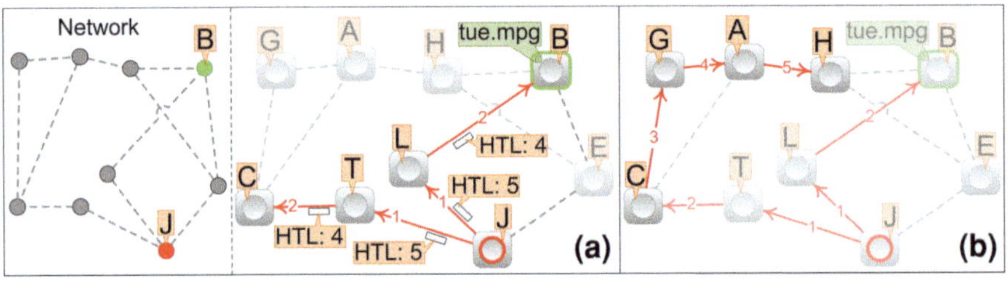

Fig. 9.5 Two random walks initiated with HTL = 5. **a** This random walk hits a resource in node *B* and terminates. **b** On the other hand, this walk continues until the HTL is exhausted, but fails to discover the resource

HTL. There is a way to achieve discovery costs that only grow linearly with HTL. The request initiator picks a maximum of *k* neighbors. However, at every following iteration, only one randomly-selected neighbor continues the process. Therefore, instead of exploring on tree-like structures, we only visit a subset of *random walks* [10], as exemplified in Fig. 9.5.

Although they manage to reduce messaging, random walks tend to lead to high hop counts for the first response hit. Due to the randomness of their propagation, walks might need to visit many nodes until they randomly select a provider node. Also, if the chosen HTL threshold is small, *success rate* and *recall* may fluctuate considerably. Caching may severely affect the success rate of random walks since once a node is revisited, the walk (one out of the very few *k*) is purged.

9.5 Informed Discovery in Unstructured Networks

All blind search algorithms described above, apart from BFS, need to randomly select one or more neighbors almost at every point of a request propagation path. Instead of this random neighbor selection, *Informed search* techniques such as *Intelligent search* (IS) and *Adaptive probabilistic search* (APS) give priority to the most promising neighbors. These two mechanisms try to direct the request towards the resource based on the knowledge accumulated on previous requests.

Intelligent search [7] is based on mBFS, but introduces certain neighbor selection heuristics. Each node profiles its neighbors, keeping track of the most recent requests as well as the successful discoveries (the *hits*). Thus, upon receiving a discovery request, a node can build a ranking table for each of its neighbors and forward the request to the highest-ranked neighbors (i.e., those who have previously processed similar requests and have drawn the most responses). Two ranking factors, *similarity* and the number of *hits*, are connected with a relevance ranking function which gives the actual neighbor ranking. Figure 9.6 shows IS in action through a simple example.

Intelligent search tends to direct the queries towards those neighbors that have previously provided responses (and thus rank high). This accelerates the discovery process, but concentrates the requests among a few nodes rather than distributing the load. To address this trap, IS still selects a fraction of neighbors in a random fashion, providing that the number of random selections is smaller than the parameter *W* of mBFS.

Adaptive probabilistic search [11] tries to lower the message cost of IS and introduces an "unlearning" mechanism to help prevent looping requests. Every node maintains a performance index for each neighbor, including all the requests processed by the neighbor. Upon receiving a request, a node accesses the relevant performance index entry and derives the neighbor selection probabilities. APS is based on the random walks scheme. Nodes do not randomly relay requests, but next neighbors are picked based on the calculated selection probabilities.

The performance indexes are always updated based on the outcomes of the requests. In particular, APS employs two schemes for updating the neighbor performance indexes: *optimistic* and *pessimistic*. In the optimistic scheme, the request-specific performance index of the visited neighbors increases "proactively," while the request propagates and decreases "reactively" upon failure. The opposite operations are performed in the pessimistic case. The *increasing* and *decreasing* rates are distinct configuration parameters which do not necessarily have to be equal. In general, the selection probability of a node is derived from the ratio between the node's index and the sum of the indexes of all other neighbors. Figure 9.7 presents an example of how APS works.

Both IS and APS extend the fundamental principles of *blind search* with a learning module which helps improving success rate and recall by trying to learn where the resources actually reside. This approach works well if the network is not fast-changing (e.g., little node mobility). By contrast, under more dynamic network conditions, the process of learning becomes less precise and may even become counter productive as it involves "unlearning" the incorrect conditions before the new status can be "learned."

9.6 Discovery in Loosely-Structured Networks

The main problem with the *unstructured* discovery protocols is that they have to deal with resources that may be located anywhere in the network. On the other

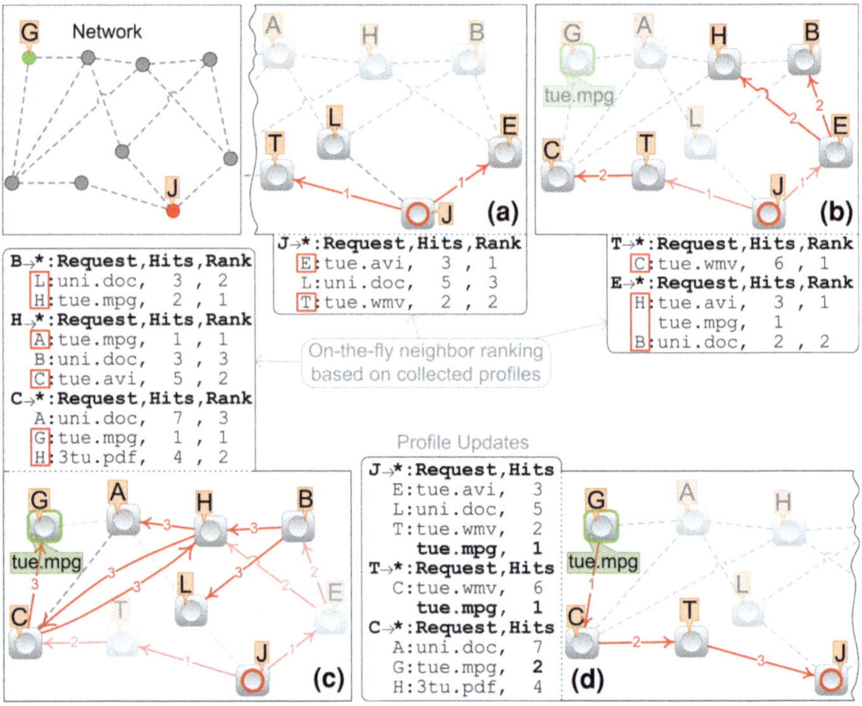

Fig. 9.6 Intelligent search based on mBFS, with $W = 2$ and HTL $= 3$, looking for "tue.mpg." **a** The request initiator ranks the neighbors and picks the best two. **b** Node T can only pick one neighbor and node E can pick both neighbors as $W = 2$. Node H has provided four hits for two different but similar requests and, hence, is ranked first among neighbors of E. **c** Subsequent nodes do the same upon receiving the request. **d** The response travels back through the path which has led to the discovery, updating the profiles of each intermediate node

hand, *structured* networks correlate the search terms with the node identifiers. Let us consider *Freenet* (http://freenetproject.org/) [8], which exemplifies a system based on a loosely-coupled relation between resource names and host identifier.

Recall from Chaps. 7 and 8 that in *Freenet,* the node identifiers are generated through the collaboration among adjacent nodes and that neighboring nodes tend to have correlated identifiers. Furthermore, each node knows a priori:

- the key with which a resource has been published into the network;
- that those nodes having an identifier "closer" to the resource key have higher probability to host the resource.

The *Freenet* discovery algorithm adopts DFS, but the neighbor is chosen based on the similarity between the *request key* and the neighbor ID (in contrast to the random choice of neighbors made by pure DFS). Thus, the search process is more efficient than in random DFS, as can be seen in the example of Fig. 9.8. However,

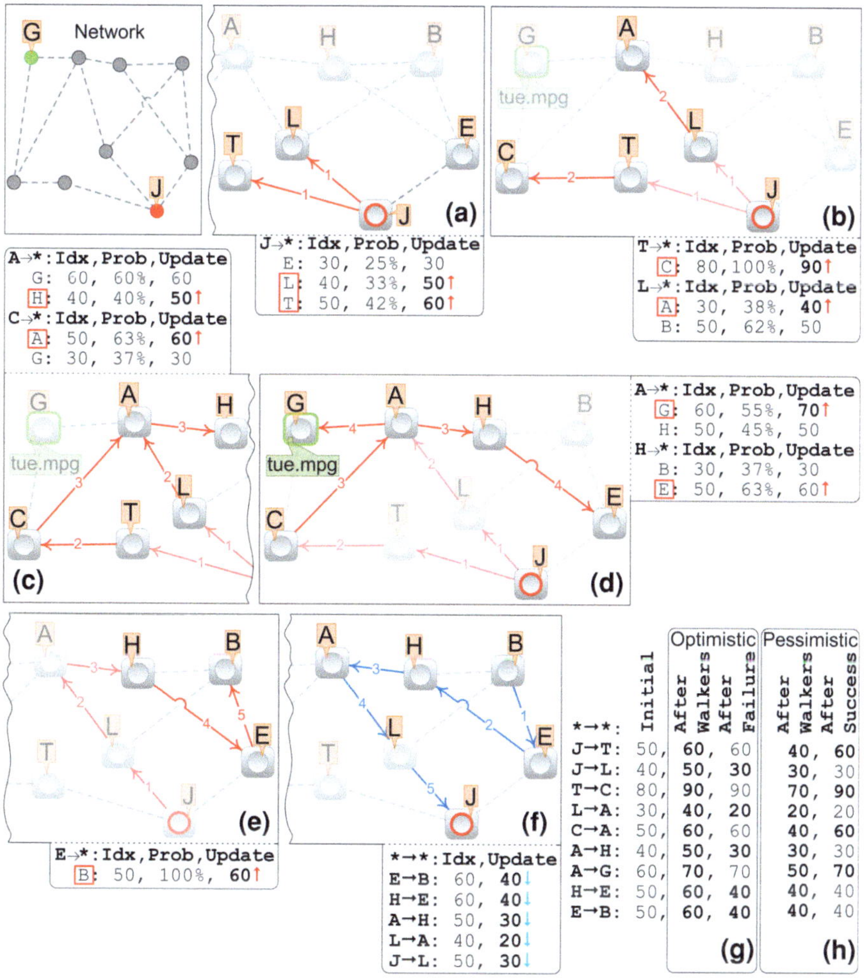

Fig. 9.7 Node J starts Optimistic APS based on 2-Random Walks, with HTL = 5 targeting a resource stored in node G. At each hop, walker nodes increase the index of the selected neighbors by ten degrees. **a** The request initiator picks the two neighbors having the highest index (nodes L & T) for the specific request with probability 40/(30 + 40 + 50) = 33% and 50/(30 + 40 + 50) = 42%, respectively. **b** Node T has no other option but C. Due to probabilistic selection, node L picks A despite the lower probability. **c–e** Similar selection procedures for the two walkers. **f** HTL expires and a failure propagates backwards, reducing the index of nodes in the path by 20°. **g** The resulting optimistic APS index tables of the nodes selected by the walkers after the walkers and after the updates. **h** The resulting pessimistic APS index tables

an issue is that as the network becomes more dynamic, the loose structuring method of *Freenet* tends to break the neighborhood of correlated node identifiers. Studies have shown that under highly dynamic conditions, the performance of *Freenet* is not much better than *Random DFS*.

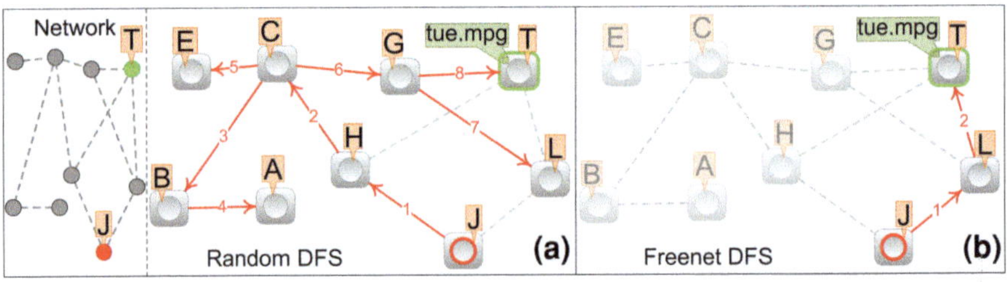

Fig. 9.8 Freenet depth-first search of key "tue.mpg" on a network with alphabetically correlated IDs of neighboring nodes. **a** A random DFS may spend many messages until it detects the resource. **b** In Freenet, the search process is directed towards the nodes whose identifiers are "closer" to the file name. Nodes "L" and "T" are more similar to "tue.mpg" than "H" and "G"

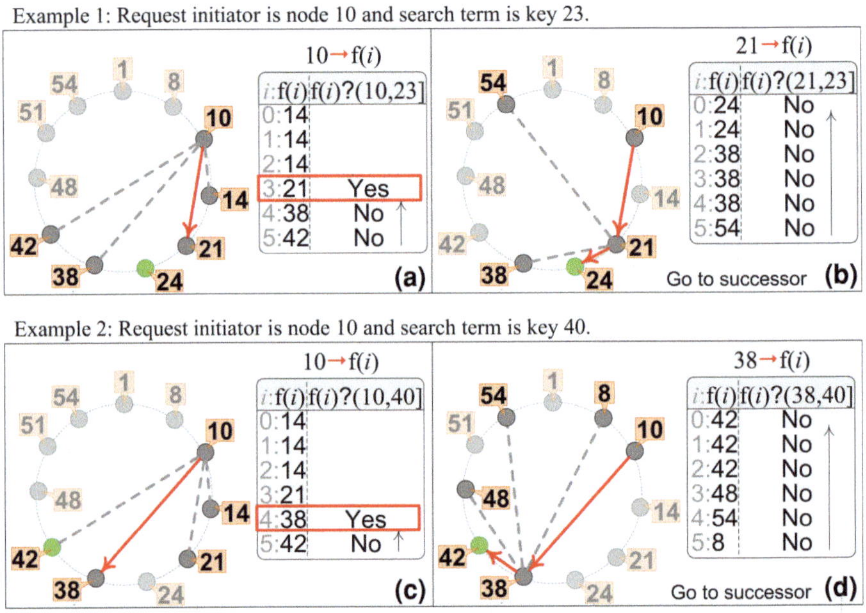

Fig. 9.9 A *Chord* look-up triggered by node 10, looking for resource keys 23 and 40. **a** The forth finger ($i = 3$, node 21) of the request initiator is the first from the bottom to be located between the current node (10) and target node (23), i.e., $21 \in (10,23]$. **b** Node 21 has no finger located between 21 and 23; thus, the immediate successor hosts the resource. **c**, **d** The request initiator picks the fifth finger as node $38 \in (10,40]$; node 38 forwards the request to its immediate successor

9.7 Deterministic Discovery in Structured Networks

While the search algorithms of *unstructured* networks target the resources, the discovery mechanisms on *structured* systems seek specific node identifiers. This is because resources are indexed or hosted by specific IDs. Any request initiator has

Example 3: Request initiator is node 10 and search term is key 0.

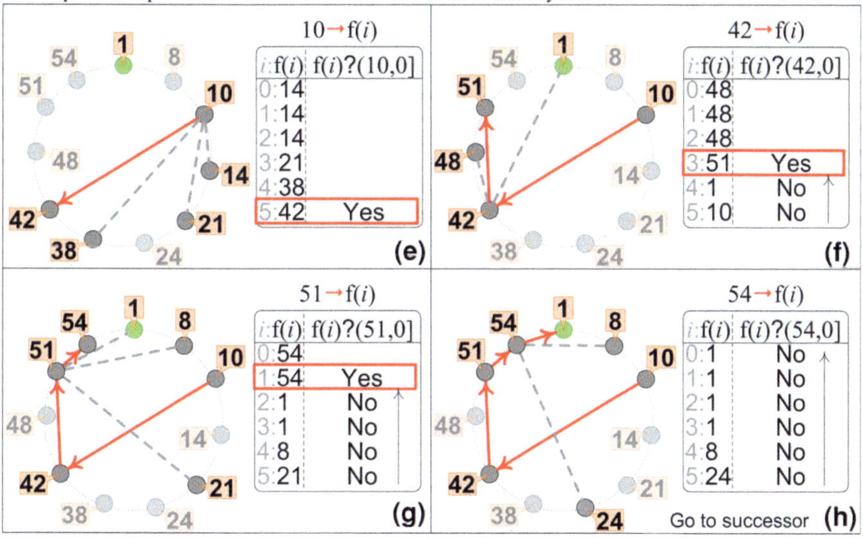

Fig. 9.10 A *Chord* look-up triggered by node 10, looking for the resource key 0. **a** Node 42 is in the part of the ring between keys 10 and 0, clockwise. **b** Node 51 is between 42 and 0. **c** Node 54 is located between 51 and 0. **d** Node 54 is the predecessor of the target node 1

enough knowledge to calculate the host ID of the requested resource before the request is forwarded. In structured networks, the discovery mechanisms only build a single propagation path, without setting any stopping criterion other than a hit on the appropriate node.

Structured networks, for example, the P2P systems based on *distributed hash tables* (see Chap. 8 for *Chord*), order the nodes in increasing ID [3–5]. The request has to be forwarded from the initiator to the target node in such a way that it will not overpass the target. In DHTs, a requesting node generates the hash value to be used as the search term—the resource key. The key is essentially produced by the same function that generates node IDs in the network; thus, the hash value is taken from the index space of node identifiers. The key is also the node identifier that most likely hosts the resource.

Let us consider, for example, the discovery process in *Chord* [12, 13]. Figures 9.9 and 9.10 include the three cases in which node 10 is looking for resources 23, 40 and 0, respectively. Nodes are sorted on a list, based on their identifier (each node holds a portion of the list). A request initiator sends the request as closely as possible to the target node, but without overshooting. A node which is either initiating or simply relaying a request picks the largest possible neighbor ID (but no larger than the target node ID). The node checks its finger table from the bottom upwards and selects the first finger that has an ID larger than the current node, but smaller or equal to the target node ID. If no such finger can be found in the node's finger table, then the immediate on hop successor is actually the host of the requested key.

The *Chord* look-up algorithm belongs to a family of mechanisms that guarantee discovery, provided that the resource exists in the network. Moreover, the *recall* is also 100% as all the resources that satisfy the request term (hash key value) are hosted or indexed by the same node. In terms of average *hop-count*, *Chord* can ensure an upper bound of $\lceil \log N \rceil$, whereby N is the size of the network. As the forwarding scheme is 1-walker, the message cost of *Chord* coincides with the worst case *hop-count*. The main limitation of structured protocols such as *Chord* is that they may lead to extensive discovery paths and, therefore, large response times.

References

1. Vu QH et al (2009) Peer-to-peer computing: principles and applications. Springer, Heidelberg
2. Zeinalipour-Yazti D et al (2004) Information retrieval techniques for peer-to-peer networks. Comput Sci Eng 6:20–26
3. Lua EK et al (2005) A survey and comparison of peer-to-peer overlay network schemes. IEEE Commun Surv Tutor 7:72–93
4. Meshkova E et al (2008) A survey on resource discovery mechanisms, peer-to-peer and service discovery frameworks. Comp Netw 52:2097–2128
5. Androutsellis-Theotokis S, Spinellis D (2004) A survey of peer-to-peer content distribution technologies. ACM Comput Surv 36:335–371
6. Chawathe Y et al (2003) Making gnutella-like P2P systems scalable. In: Conference on applications, technologies, architectures, and protocols for computer communications, ACM, pp 407–418
7. Kalogeraki V et al (2002) A local search mechanism for peer-to-peer networks. International conference on Information and knowledge management, ACM, pp 300–307
8. Clarke I et al (2001) Freenet: a distributed anonymous information storage and retrieval system. In: Federrath H (ed) Designing privacy enhancing technologies. Springer, Heidelberg, pp 46-66
9. Reinefeld A, Marsland T (1994) Enhanced iterative-deepening search. IEEE Trans Pattern Anal Mach Intell 16:701–710
10. Gkantsidis C et al (2006) Random walks in peer-to-peer networks: algorithms and evaluation. Perform Eval 63:241–263
11. Tsoumakos D, Roussopoulos N (2003) Adaptive probabilistic search for peer-to-peer networks. International conference on peer-to-peer computing, pp 102–109
12. Stoica I et al (2003) Chord: a scalable peer-to-peer lookup protocol for Internet applications. IEEE/ACM Trans Netw 11:17–32
13. Karrels D et al (2009) Structured P2P technologies for distributed command and control. Peer-to-Peer Networking and Applications. 2:311–333

Chapter 10
A Peek at the Future Internet

Abstract The Internet "connectivity machine" is the generative engine of our modern digital society. It has been the launching pad of the Web (now the Web 2.0), truly the largest and most versatile information system ever built. While the Web phenomenon relentlessly continues, scientists worldwide are now living the dream of yet a more generative next-generation network. This chapter explores some prominent research directions, discussing the *Internet of Things*, *context-aware networks*, *small world networks*, *scale-free networks*, *autonomic networks*, *dependable networks*, the *privacy vs. security* dichotomy and the two facets of *energy-efficient networks*.

The best way to predict the future is to invent it
Alan Key, computer scientist

10.1 The Fourth Networking Principle: Beyond Mere Connectivity

In Chap. 2, we introduced the three fundamental principles of networking: *connect*, *discover* and *stay connected*. It is now time to argue that the next-generation networks should go beyond mere connectivity.

Networks are currently engineered in layers, going from the lower *physical* layer up to the *application* (the seventh) layer. Each layer has specific responsibilities (for instance, layer three, the network layer, is in charge of *route computation*) and dedicated interfaces with the adjacent layers. This "insulation" between layers makes networks more manageable and facilitates the appearance of new applications. Layers make it easy to focus on specific functionalities without having to

A. Liotta and G. Exarchakos, *Networks for Pervasive Services*, Lecture Notes in Electrical Engineering, 92, DOI: 10.1007/978-94-007-1473-1_10, worry about the whole system. In fact, the marvelous Internet applications that we use today are probably the direct consequence of the layering model: the programmers could focus on the application logic without having to master the lower layers.

However, inter-layer insulation comes with a downside: it limits our ability to introduce new optimization mechanisms. If we keep the network layer isolated, the routing and transport functions cannot promptly take into account the requirements arising from the physical network. What is worse, *packet routing* does not adapt to the applications or to the user's context. For instance, we cannot implement content-based routing on the IP layer.

The advances made at the network edges, such as P2P networking, have revealed the potential of *context-aware* networking. At the same time, the realization of context-aware networks at the application level (as in P2P) is not ideal. The damage that P2P applications cause to the network is now well documented. On the other hand, IP networks are not always capable of meeting the delivery deadlines of the real-time P2P systems (such as P2P IPTV). In fact, we can now observe a trend whereby *cloud* services such as *YouTube* are becoming more prominent.

Imagine what we could achieve if the network itself could take into account requirements and constraints arriving from the other layers. We could route packets based on the type of content, the user's context or the recipient's preferences. The network would be able to spot communication patterns and allocate resources accordingly. Routing algorithms would be based on a *probabilistic* approach rather than on the current *deterministic* approaches that do not work under dynamic conditions.

Context-awareness is the extra gear that is missing in the Internet and a crucial mechanism for the realization of the next-generation Net. The upcoming networks will not be completely autonomic,[1] but will certainly have to be more adaptive regarding a variety of perturbations.

10.2 Internet of Things: Sense and Influence Your Environment

The time when the Internet was for the sole use of computers is over. Our technology roadmap is going towards the *Internet of Things* (IoT) [1, 2], a digital infrastructure where anything having any kind of network interface will be part of the Net. The convergence between the conventional stationary Internet and the cellular network has given tremendous impulse to the digital society [3]. Even greater breakthroughs will come from the interconnection of everyday objects, sensors and actuators.

[1] Autonomic networks are envisioned to be able to *self-configure*, *self-heal*, *self-optimize* and *self-protect* with minimal human intervention, according to the autonomic computing principles.

The realization of the IoT poses ambitious scientific hurdles, though it certainly has enormous potential. With virtually anything on the Net, from the domestic appliances to clothing and biometric sensors, the network will suddenly assume a "massive" scale.

Yet the biggest challenge will probably come from the huge functional diversity among the devices. RFIDs[2] can do very little in terms of networking, but give a cheap way to locate a myriad of objects. Multiple sensors may collaborate to provide environmental monitoring information, but will have substantial computational and energy constraints. Intelligent camera systems may solve complex surveillance problems, though they will incur severe traffic onto the network.

The size and diversity of the IoT cannot be handled by the current IP protocol [4]. On the other hand, the IoT will be able to rely on a wealth of contextual information that will enable greater routing intelligence. The IoT will not only propagate contextual information "where" and "when" it is needed, but it will also make use of the context to better operate the network itself.

Once we make the move to attaching anything to the Net, the network will become the largest control system ever built. The network's "things" will provide *sensory*, but also *transducing* and *actuation* capabilities. Actuators, for example, motors, pneumatics and hydraulics, can move objects and pump fluids. Electrical relays can switch on the heating system or turn off the lights.

The transducers will further enhance the network's self-sufficiency. Researchers are making progress in the area of energy-harvesting transducers that can capture small but usable amounts of energy from the environment. This energy can be used to run sensors and network interfaces.

The next-generation network will be able to *grasp* and simultaneously *influence* its environment. Scientists are investigating the paradigm shift required to make the most of these new capabilities.

10.3 Small, Large Networks

There is no doubt that the Net is getting bigger, more complex and increasingly dynamic. At the same time, the perturbations created by emerging applications are more and more intense and erratic. The Net is a complex system that is constantly changing and expanding. The routing protocols must keep everything connected; they must discover short paths across such a massive network.

One way to keep large networks "small" is to increase the number of links, making the network denser. This is easier said than done. Adding new capacity on

[2] Radio frequency identification (RFID) is a technology that uses communication via radio waves to exchange data between a reader and an electronic tag attached to an object for the purpose of identification and tracking.

the physical network is costly. In fact, the current Net is relatively sparse; it has a number of links roughly of the same order of magnitude as the number of nodes.

Things get more complicated if we try scaling up the network while at the same time ensuring "stability." Suppose we can add new links. How do we know which node pairs would benefit the most from the extra capacity? Where do we add capacity in a constantly changing network? How can we make this choice automatically?

Ironically, while the *computer networks* community has created a marvelous yet complex digital ecosystem, fundamental breakthroughs have also been achieved beyond the technologists' circle. Physicists, biologists, mathematicians and sociologists have been studying biological [5] and neural networks [6] that are far more complex than the present Internet [7, 8]. Thus, understanding the properties of the "natural" networks should be the starting point for those who are rethinking the Internet [9–11].

Perhaps one of the most remarkable discoveries is the *small-world* phenomenon, which is present in most complex networks [7]. Apparently, the networks resulting from a natural evolution process are able to build short paths, irrespective of the number of nodes. A fascinating yet not fully proved theory is that in natural networks, any node is, on average, six hops away from any other node—this is known as the "six degrees of separation" property or "small-worldness".

Another outstanding property of natural networks is known as *scale-freeness* [7]. Scale-free networks exhibit the same interconnectivity distribution, no matter how big the network grows. While *small-worldness* is key to scalability, *scale-freeness* is crucial for robustness and stability.

The mechanics of small-world and scale-free networks is not fully understood. However, scientists have already unveiled several mysteries. We have enough knowledge to start designing routing protocols that can make a large network "small."[3] We know that a well-designed network must have short paths. This can be achieved if the network has the "right" mixture of low- and high-degree nodes and of weak and strong links [12].[4]

Scientists have discovered a number of counter-intuitive properties that have significant potential for the re-design of routing protocols. For instance, weak links play a crucial role in reducing the network diameter as they build long-distance bridges between nodes that would otherwise be poorly connected. Because of their nature, weak links tend to be transient. It seems to defy logic, but scientists have discovered that it is precisely this volatility that makes weak links so crucial in kicking the network out of sub-optimal configurations. Weak links make it possible to propagate signaling information more rapidly and towards areas that would

[3] Recent literature describing the properties and mechanisms of small-world and scale-free networks is included in our "References" section.

[4] A link is "weak" when its addition or removal does not significantly change the mean value of a target measure (P. Csermely, "Weak Links", Springer 2009).

otherwise not be reached. Weak links hold the secret of *stability*. However, weak links cannot exist without the strong ones. In fact, the natural networks have a continuous spectrum of link strengths.

Extensive studies of complex networks have unveiled how difficult it is to pursue multiple performance goals. Network *speed* and *stability* are often conflicting targets. It is a myth that networks' diameter can be merely reduced by increasing the average node degree. Nodes with a large number of neighbors are called *hubs*. Hubs multiplex traffic; so they are important. However, hubs come with a problematic side effect. They make the network vulnerable. Hubs have huge responsibilities; so if they are attacked, large portions of the network are affected. Hubs not only propagate genuine data, but also speed up the spreading of computer viruses or any other destabilizing agent.

Ironically, hubs and strong links help to improve transmission speed, but do not play a positive role when it comes to stability and robustness. Another counter-intuitive finding is that in addition to weak links, *bottlenecks* can also help make networks more *robust*. Bottlenecks limit the network throughout, but often generate new weak links. Bottlenecks force networks to re-distribute the load and trigger a rewiring process that is crucial in protecting networks against cascading failures. Scientists such as Motter have proved that a selective removal of network elements makes the network more robust.[5]

One of the problems of the current routing protocols is that they strive for a "uniform" network. They pursue routing efficiency but neglect other essential properties. Looking at the most complex natural networks, we see that they are not only transmission-efficient, but also tolerant to incredible amounts of failures, errors, noise and dynamics. Small-world, scale-free networks have a mix of randomness, nestedness,[6] disuniformity, volatility and unpredictability. They have a variety of nodes (hubs,[7] rich clubs,[8] VIP clubs,[9] leaves and bottlenecks) and links (bridges, weak and strong links). As part of their evolution, the natural networks have learned how to orchestrate this variety of elements and respond to new forms of perturbations.

---

[5] A.E. Motter, Cascade control and defense in complex networks. Phys Rev Lett 93, 098701.

[6] *Nestedness* indicates the hierarchical structure of networks. Each element of the top network usually consists of an entire network of elements at the lower level. Nestedness helps us to explain the complexity of networks.

[7] *Hubs* are connection-rich network elements.

[8] In hierarchical networks, the inner core becomes a *rich club* if it is formed by the hubs of the network. For example, in the Internet, the routers form rich clubs.

[9] In VIP clubs, the most influential members have low number of connections. However, many of these connections lead to hubs.

10.4 Manage the Autonomics

Networks are becoming increasingly complex and heterogeneous. Networks are nested within networks, virtualized, overlaid, sub-netted. Some sections of the Internet are "managed," e.g., by network operators or ISPs. However, there is a steep increase in "unmanaged" networks (e.g., wireless home networks), "spontaneous" networks (e.g., *ad hoc* networks) and "content-driven" networks (e.g., P2P networks). Several researchers are investigating how to bring the power of the natural evolutionary networks into the Net [5, 6, 13]. By mimicking biological mechanisms, the "bio-inspired" computer networks promise efficiency, robustness, scalability, but also "adaptivity" and "evolvability."

In the future, big chunks of the Net will be "autonomic" [3, 14]. Networks will be able to learn how to respond to new kinds of perturbations. They will be able to absorb and disperse the bad signals whilst transmitting the good ones. They will be resilient to viruses, failure or catastrophic events.

Many networks will be self-managed [2, 15], though human intervention will still be needed. It will be necessary to incorporate higher-level management mechanisms to manage the complex entangle of autonomic elements. There is a possibility that the introduction of sophisticated automatisms will generate new problems in terms of signaling, stability, security and trust. The multiplicity of autonomic systems will interact, influencing each other. How can we ensure that such interactions do not degenerate or create interferences or instabilities?

Just as in the evolutionary networks within nature, the different sub-systems of the future Internet will morph over time. However, computer networks are influenced by multiple factors that we have not yet learnt how to master. The evolution of the Net is affected in different ways by technology, but also by economic, political, legal and social elements. Until we find out how to realize a self-sustained digital ecosystem, we shall continue to need human intervention for purposes such as global optimization, regulatory obligations, law enforcement, business and provision of quality levels [16, 17]. Thus, for many years to come, it will still be necessary to monitor the autonomics and possess a means to influence it positively.

10.5 Dependable Networks

As the Net is used more and more for time-constrained applications, we are left to deal with a critical question: how *reliable* is the Net? We mentioned earlier that the typical packet-loss rate is in the order of 8–10% (internettrafficreport.com). However, even though so many packets are dropped, we can still run a variety of applications [18]. This has been made possible by innovating the applications rather than trying to introduce better network mechanisms. The innovation has taken place on the network's edges through techniques such as *caching*, *adaptive coding*, *scalable coding* or *P2P transmission* (to mention just a few) [19–24].

Yet, the transition from the current *best-effort* network to a more *dependable* Net will be unavoidable if the unrelenting trend towards extreme *ubiquity* and *mobility* continues. At present, networks put little effort into delivering packets on time. When a packet is dropped, the IP layer has the tendency to forget about it and move on with normal life. This means that not much is done in the core network, apart from buffering the packets during congestion periods. However, *buffering* is not the ultimate solution. It is just a temporary patch that has the ability to deal with transient problems. The very heart of the network, the all-optical trunks, is not even able to perform any buffering.[10] Also, by buffering a packet, we shield it from congestion while, at the same time, incurring extra latency.

The existing network mechanisms concerned with *congestion* and *packet loss* are rudimentary. Packets are dropped unpredictably. When a loss is detected at the application layer (e.g., through TCP or within the application), we can try to recover the packet by requesting a retransmission. Yet, just like buffering, *retransmission* affects the communication latency. What is worse, there is no way to determine whether the retransmitted data can in fact meet its delivery deadline. Obviously, a "failed" retransmitted packet (one that reaches the application too late and is unusable) incurs unnecessary traffic. Also, if a packet is dropped due to congestion, chances are that the retransmitted packet will suffer the same destiny. Thus, packet retransmission, as such, does not provide a solid answer to "reliable" transmission and is potentially a counter-productive measure (retransmissions worsen congestion).

The key reason why the Net does not offer robust transmission is lack of parallelism. If a path is congested, the Net tries to find a diversion, and it does so pretty slowly. The lesson learned from P2P applications is that a much better approach is to seek alternative sources. There is huge data redundancy in the Net; therefore, it does not make sense to only transmit via point-to-point channels. Exemplar transmission mechanisms are *chunk-based* transport, *multi-layered* transport, *redundant-data* transport and *location-depended* transport such as in cloud services.

The foundations of dependable networks lie on intelligent routing and transport mechanisms that incorporate *parallelism*, *context-awareness* and *content-awareness*. Significant research efforts are currently directed towards such a *cross-layer routing* approach [25].

10.6 The Fine Line Between Freedom, Security and Privacy

Many people today trust the Net with their most private data. To make a transaction, we post our credit card details and other personal data. While we browse through an e-shop, we actually leave traces of our preferences. Blogging and twitting are vehicles for people to express opinions, which, in some countries, may

[10] Optical buffering is currently one of the major hurdles in the realization of all-optical networks, which would lead to a substantial increase in network capacity.

lead to persecution or prosecution. On the other hand, monitoring, tracing and logging over the Net are required for the purposes of *law enforcement* and *cybercrime prevention* [26].

Thus, a most controversial issue is how to balance freedom, security and privacy [27]. A number of technical as well as legal tools have been developed for the purpose of (and in the name of) security. For instance, telecom providers must be able to supply so called *legal intercepts* in response to a court order. Similarly, ISPs are required to disclose to the judge the personal whereabouts of alleged cybercriminals.

However, within the Net, there is only a very fine line between security, crime and prosecution: the very same security tools may be misused by cybercriminals or by totalitarian regimes [28].

The Net is not well equipped to protect our privacy. Thus, a number of techniques are being developed to tackle the issue in a radical way, for example, through anonymity-support services [29, 30]. Clearly, if the identity of a blogger is hidden to the ISP, he can voice his opinion online without fear of being persecuted. Yet again, anonymity works against security because it allows cybercriminals to hide.

Looking at the present technology, there is no apparent solution to the "freedom-privacy-security" clash. This is possibly because the Net, that is, the primary handler of sensitive data, is totally oblivious to the problem. The IP is geared for "sharing" more than it is for "securing" data. It comes with some raw mechanisms for encrypting packets, but is oblivious to the issues of *freedom*, *privacy*, and *security*. Unlike any other complex network (e.g., the biological networks), the Net has no self-defense mechanisms. The result is that a predominant fraction of the Internet traffic today relates to *viruses*, *spam*, *polluted content*, and *malicious software*. The Net does not know how to identify and slow down the propagation of "bad" data whilst accelerating the distribution of the "good" one.

The consequences are dramatic. An estimated 80% of emails are spam, which costs businesses in the range of \$20.5 billion annually, a figure that will soon rise to \$200 billion (spamlaws.com). Identity theft hits around \$10 million Americans a year and costs businesses about \$200 billion a year. Similarly astonishing figures relate to viruses and the other plagues of the Net.

Despite the significant attention by the policy-makers, the critical issues surrounding freedom, privacy and security are still partly in the hands of the cybercriminals. This complex issue must be tackled at a global scale and in every element of the digital society. The next-generation network can no longer remain out of the problem.

10.7 Energy-Efficient Networks

Harvard scientists estimate that today "the whole ICT sector produces as much greenhouse gases as all airline companies together" [31]. An increasing fraction of energy is consumed by the large data centers—1000 *Google* searches burn about as much energy as driving 1 km with a car. The network itself takes a large toll. A 2007 study by *Telecom Italia* unveiled that its network absorbed over 2TWh, representing 1% of the total national demand, which ranked the company as the second largest energy consumer (after the *National Railways*).[11]

As much as they have become crucial to the global economy, today's networks are not at all eco-friendly. They are always ON and burn energy even when they are on standby. The energy consumption of an Internet connection is dominated by the access network, particularly the broadband access lines. The deployment of fiber-to-the-home (FTTH) will lead to substantial improvements (optical transmission is more efficient than electrical). Substantial effort is in fact directed towards all-optical networks, but many fundamental issues still remain open: how can we build all-optical routers if we do not know how to construct suitable optical buffers?

In parallel to the research efforts on the "physics" of networks, significant breakthroughs are required on the "soft" aspects. Improving the routing architectures along the lines indicated earlier in this chapter (*context-awareness*, *small-worldness*, *autonomicity*, *parallelism* etc.) will be a priority.

Another dimension of energy-efficiency [32] is created by the convergence between the conventional networks and the emerging variety of wireless networks. These span beyond the confines of WiFi, WiMax and the cellular networks. Spontaneous, opportunistic connectivity along the concepts of *ad hoc networks* and *IoT* will surely gain importance. In this context, energy-efficiency is required at a different level, not just to save the planet. The emerging edge-network will comprise a range of battery-operated terminals that will participate in the complex *routing* game. Terminals will *source*, *filter*, *clean*, *store*, *relay* and *terminate* data. The terminals will play a key role in differentiating between "good" and "bad" data, filtering spam, eradicating viruses, aggregating data and help to propagate it reliably. In this way, the "edge" networks will help to keep the traffic local; they will have the ability to spot communication patterns, self-regulate and minimize their energy consumption. This vision of *cognitive networking* is debated passionately at the present, although we do not yet know how to realize autonomic networks that are also controllable, stable and reliable.

10.8 No Matter What, the Network will Remain Generative

We started this book writing about the "generative" power of the Internet [33]. On its conception, the Internet had not been designed in view of the phenomenal applications it is still sparking. Today, the search giant *Google* is valued at $200

[11] C. Bianco, F. Cucchietti, and G. Gri, "Energy consumption trends in the next generation access network—a telco perspective," International Telecommunications Energy Conference, INTELEC, Sep. 30–Oct. 4, Rome, Italy, 2007.

billion and the social network *Facebook* is worth $50 billion.[12] Yet neither application was in the minds of the Internet architects, Robert Kahn and Vint Cerf.

Along with *creativity,* comes the urge to protect it from spam, viruses, computer hackers and the lot. Many attempts to protect our digital assets have resulted in measures that have, at times, restrained the "generativity" of the Net. For instance, we have seen a periodic alternation between those who support an "open" networked environment (one that gives wide freedom to individuals to manipulate their terminals, as in *Linux*) and those who are prepared to sacrifice this freedom in exchange for a greater sense of security (as in the video game consoles which do not allow any customization by the customer).

Nevertheless, even this emerging form of an "appliancized" network has not actually stopped the generativity of the Net. The iPhone/iPad phenomenon has proved that *device tethering* does not always confine our inventiveness. The apple store had 50,000 applications available in 2009 and 330,000 in January 2011, with a rate of 600 new applications submitted every day.[13] In comparison with other open platforms such as *Google Android*, we observe a counter-intuitive phenomenon: the sense of security instilled by a *tethered* network does occasionally surpass the sense of freedom instilled by an *open* network.[14]

Despite the unexpected developments of the last decade, the Internet "connectivity machine" has not seen many changes since its conception. Yet, it has continued to be the "generative" engine of our digital society. Scientists worldwide are now living the dream of yet a more generative next-generation network.

References

1. Chaouchi H (2010) The internet of things: connecting objects. Wiley-ISTE
2. Kim S, Choi M, Ju H, Ejiri M, Hong J (2008) Towards management requirements of future internet. Challenges for next generation network operations and service management. Springer, pp 156–166
3. Liotta A, Liotta A (2008) P2P in a regulated environment: challenges and opportunities for the operator. BT Technol J 26:150–157
4. Akyildiz IF, Vuran MC (2010) Wireless sensor networks. Wiley
5. Farooq M (2010) Bee-inspired protocol engineering: from nature to networks. Springer, Heidelberg
6. Haykin S (2008) Neural networks and learning machines. Prentice Hall, New Jersey
7. XiaoFan W, Guanrong C (2003) Complex networks: small-world, scale-free and beyond. IEEE Circuits Syst Mag 3:6–20
8. Dressler F, Akan O (2010) Bio-inspired networking: from theory to practice. IEEE Commun Mag 48:176–183

---

[12] *Source* The Economist, 8–14 January 2011.

[13] *Source* http://148apps.biz/app-store-metrics/

[14] As of January 2011, the number of *android* devices has surpassed the number of iPhones. However, the number of iPhone apps is significantly larger.

9. Buchanan M (2003) Nexus: small worlds and the groundbreaking theory of networks. W.W. Norton & Company
10. Gammon K (2010) Four ways to reinvent the Internet. doi:10.1038/463602a
11. Clark DD, Partridge C, Braden RT, Davie B, Floyd S, Jacobson V, Katabi D, Minshall G, Ramakrishnan KK, Roscoe T, Stoica I, Wroclawski J, Zhang L (2005) Making the world (of communications) a different place. SIGCOMM Comput Commun Rev 35:91–96
12. Csermely P (2009) Weak links: the universal key to the stability of networks and complex systems. Springer
13. Dhillon S (2008) Ant routing, searching and topology estimation algorithms for ad hoc networks. Delft University Press
14. Papadimitriou D (ed) (2009) Future internet—the cross-ETP vision document, v. 1.0. http://www.futureinternet.eu
15. Jennings B, van der Meer S, Balasubramaniam S, Botvich D, Foghlu M, Donnelly W, Strassner J (2007) Towards autonomic management of communications networks. IEEE Commun Mag 45:112–121
16. Agboma F, Liotta A (2010) Quality of experience management in mobile content delivery systems. J Telecommun Syst (Spacial issue on the Quality of Experience issues in Multimedia Provision, 14 pages, Springer; Online First, 24 June 2010). doi: 10.1007/s11235-010-9355-6
17. Menkovski V, Exarchakos G, Liotta A, Sánchez AC (2010) Quality of experience models for multimedia streaming. Int J Mobile Comput Multimedia Commun (IJMCMC) 2(4):1–20. doi: 10.4018/jmcmc.2010100101
18. Handley M (2006) Why the Internet only just works. BT Technol J 24:119–129
19. Alhaisoni M, Liotta A (2009) Characterization of signaling and traffic in Joost. Peer-to-peer Netw Appl 2:75–83
20. Alhaisoni M, Liotta A, Ghanbari M (2009) Improving P2P streaming methods for IPTV. Int J Adv Intell Syst 2:354–365
21. Alhaisoni M, Ghanbari M, Liotta A (2010) Scalable P2P video streaming. Int J Bus Data Commun Netw 6:49–65
22. Alhaisoni M, Ghanbari M, Liotta A (2010) Localized multistreams for P2P streaming. Int J Digit Multimed Broadcasting 2010:1–13
23. Alhaisoni M, Liotta A, Ghanbari M (2010) Resource-awareness and trade-off optimization in P2P video streaming. Int J Adv Media Commun 41:59–77 (special issue on High-Quality Multimedia Streaming in P2P Environments)
24. Alhaisoni M, Liotta A, Ghanbari M (2010) Resilient P2P streaming. Int J Adv Netw Serv 3:209–219
25. Iannone L (2009) Cross-layer routing and mobility management in wireless mesh networks. VDM Verlag
26. Liotta A, Liotta A (2011) Privacy in pervasive systems: legal framework and regulatory challenges. pervasive computing and communications design and deployment: technologies trends and applications. Idea Group Publishing
27. Mackinnon R (2010) Liberty or safety? Both-or neither (spectral lines). IEEE Spectr 47:10
28. Rennhard M (2004) MorphMix—a peer to peer based system for anonymous internet access. Shaker Verlag GmbH, Germany
29. Leavitt N (2009) Anonymization technology takes a high profile. Computer 42:15–18
30. Clarke I, Miller S, Hong T, Sandberg O, Wiley B (2002) Protecting free expression online with Freenet. IEEE Internet Comput 6:40–49
31. Bianco C, Cucchietti F, Griffa G (2007) Energy consumption trends in the next generation access network—a telco perspective. Int Telecommun Energy Conf 737–742
32. Restrepo J, Gruber C, Machuca C (2009) Energy profile aware routing. IEEE Int Conf Commun 1–5
33. Zittrain J (2009) The future of the internet and how to stop it. Yale University Press, London

Index

A
Ad hoc networks
  ad hoc on-demand distance vector, 51, 97
  broadcast discovery algorithm, 59
  broadcast discovery benefits, 59
  broadcast discovery examples, 61
  broadcast discovery overheads, 62
  fisheye routing protocol, 88
  hybrid, 89, 90, 92, 91
  hybrid approach, 89
  optimized link state routing, 80
  path discovery, 58
  proactive approach, 79, 88
  reactive approach, 65, 75
  zone routing protocol, 90
Ad hoc on-demand distance vector
  direct path, 65, 65, 69, 70, 72, 77, 77
  dynamic path updating, 68
  error management, 74
  node synchronization, 73
  path discovery, 68
  path expansion, 72
  path management, 70
  reverse path, 69
  route error message, 75
  route reply, 69
  route request, 69, 70
  routing table, 70
  time stamping, 73
Adaptive coding, 150
Adaptive probabilistic search, 138, 139
Alphabelt, 121, 122, 125, 125, 126
Always-on-everywhere, 10
Announcement path, *see* Depth-first search, 117, 119
Anycast network, 106
Asynchronous applications, 2
Asynchronous interaction pattern, 15
Autonomic networks, 51, 149

B
Best-effort network, 2, 20, 150
Bio-inspired networks, 149
BitTorrent, 15
Bootstrap server
  decoupled from overlay, 99
  dedicated, 17, 42, 48, 58, 63, 76, 91, 98, 139
  discovery of, 118, 121, 129
  multiple, 13, 45
Bootstrapping process
  distance vector routing, 22
  node ID assignment, 112–116, 121
  proactive network, 76
  reactive networks, 73
Bottleneck, 45
Breadth-first search, 129

C
Cached Ping Pong, 111
Carrier sensing, 56
Chord, 120, 121, 123
  finger table, 139
  look up mechanism, 121
Chunk-based transport, 145
Churn rate, 119
Classless inter-domain routing, 39
Clear-to-send, 55
Client–server, 4, 5, 7, 16, 17
Communication patterns, 15
Congestion control, 40

C (*cont.*)
Connect. *See* Fundamental networking principles
Content caching, 128
Content-aware network, 91, 104
Control processing unit, xvii, 134
Cross-layer routing, 149

D
Data agnosticism, 36
Deep data discovery, 48
Depth-first search, 129, 130, 136
Destination address, 20
Differentiated services, 39
Dijkstra E.W. algorithm. *See* Shortest-path algorithm
Disaster management, 42
Discovery mechanisms, 105, 127, 128
  blind, 127, 128
  loosely structured, 127
  structured, 127
  unstructured, 127
Discovery performance
  adaptive probabilistic search, 132
  bredth-first search, 129
  broadcast-based, 60
  chord, 123
  depth-first search, 129
  intelligent search, 132
  iterative deepening search, 129
  metrics, 86
  modified breadth-first search, 129
  random walks, 129
  tree-exploration algorithms, 132
Disruptive applications, 37
Distance vector routing, 22, 42
  bootstrapping, 52
  count-to-infinity problem, 27
  robustness, 113
  table creation, 23
  table update, 23, 26
Distributed hash table, 120, 125
Domain name system, 38
Dynamic channel reservation, 55
Dynamic host configuration protocol, 33

E
Edge buffering, 3
Emergent applications, 8
End-to-end delay, 4
Energy-efficient networks, 146
Evolution of internet applications, 3
Exposed-terminal problem, 54
Extensible protocol, 38

F
Fiber-to-the-home, 146
Finger table, 121
Fisheye routing protocol, 85
Flooding, 67, 76, 80, 108
Fourth networking principle, 139
Freenet, 149, 105, 125
Full mesh network, 119
Fundamental networking principles
  connect, 35
  context-aware routing, 12
  discover, 35
  stay connected, 35
Future internet, 14, 35, 48, 89, 139

G
Garbage collection, 68
Generative network, 2, 3, 5
Gnutella, 105, 107

H
Hamiltonian cycle, 119
Hash function, 99, 115, 116, 121, 122
Hidden-terminal problem, 52, 53
Hops-to-Live, 52, 53, 54

I
ID. *See* Node identifier
Infrastructure-based network, 42
Infrastructure-less network, 42
Input buffer, 21
Integrated services, 39
Intelligent search, 132
Inter-layer insulation, 139
Internet of things, 140
Internet patching, 10
Internet protocol, 2, 34
  scanning, 101
Internet service provider, 7
Interzone routing protocol, 89
Intrazone routing protocol, 89
IPTV, 39
Iterative deepening search, 131, 138

J
Jitter, 19
Joining process, 99, 112, 117
Joost, 7, 15, 98

K
Keep it simple and stupid, 58

L
Layer three. *See* Internet protocol
Layering model. *See* Inter-layer insulation
Leaf node, 127
Link state routing, 30
Link state packet, 30
Location-depended transport, 151
Loosely-coupled applications, 2
LSP. *See* Link state routing

M
Message caching, 134
Messages spent on requests, 134
Mobile ad hoc networks, 43, 44
Mobile IP, 41
  corresponding node, 33
  foreign agent, 33
  home agent, 33
  reverse path, 33
  triangle routing, 33
Mobile streaming, 9
Modified breadth-first search, 135, 136, 139
Moving networks, 44
Multicast network, 41, 105, 107
Multi-layered transport, 151
Multiple access with collision avoidance, 56
  clear-to-send, 56
  collision scenario, 57, 58
  data packets collision, 56
  exposed-terminal scenario, 58
  hidden-terminal scenario, 57
  request-to-send, 56
  signaling packets collision, 56
Multipoint relays, 80, 80, 81, 83, 86

N
Napster, 118
Neighbor discovery protocol, 90
Neighbor list, 98, 112, 112, 115, 126
  dynamic, 100
  fixed, 100
  ranking, 138
Network
  agnosticism, 12
  anonymity, 151
  best-effort, 150
  congestion, 45
  contention, 18, 20
  convergence, 146
  dependability, 150
  dynamics, 44
  energy-efficiency, 152
  entry point, 99
  generativity, 153
  layer. *See* Internet protocol
  legal intercept, 151
  mechanisms, 13
  operator, 7
  ossification, 39, 40, 42
  patching, 3, 9
  privacy, 151
  scanning. *See* Internet protocol scanning
  security, 151
  spam, 152
  topology, 112
Network-agnostic applications, 5, 11
Network-fairness, 5
Network-friendly applications, 5
Network-neutral application, 12
Neural networks, 147
Next-generation network. *See* Future internet
Node degree, 112, 148
  incoming, 112, 127
  outgoing, 112
Node identifier, 97, 121
  centralized assignment, 98
  dynamic, 99
  static, 99

O
One laptop per child program, 43
Optimized link state routing protocol, 80
  complete example, 86, 88
  hello packet, 80
  information repository, 83
  MPR selector set, 83
  multipoint relay example, 82
  multipoint relay node, 81
  multipoint relays, 80
  packet forwarding, 83
  route computation, 83
  shortest path example, 85
  topology control messages, 80

O (*cont.*)
  topology information base, 84
Out-of-order packet delivery, 19

P
P2P paradigm, 7, 9
Packets
  packet collision problem, 53
  packet loss, 19
  packet routing, 22
  packet switching, 18
Parallel transport, 45
Peer, 95
Peer-to-peer network, 46, 95, 98, 150
Peer-to-peer TV, 6
Per-flow routing, 48
Personal computer, 49
Pervasive application, 12
Physical network, 95
Ping–Pong, 112, 114, 115, 116, 135
  cached, 115
Privacy, 41
Proactive networks, 79
  optimized link state routing, 80
Public key infrastructure, 121

Q
Quality of experience, 4, 16
Quality of service, 12, 41

R
Radio frequency identification, 146
Random walks, 135, 138, 140
Reactive networks, 65, 77
  bootstrapping of, 67
  neighbor discovery, 67
Recall of resource instances, 134
Redundant-data transport, 151
Reliable transport, 40
Rendez-vous point, 118
Request hit, 138
Request-to-send, 56
Resource adverts, 118
  advertisement boards, 118
  multiple replicas, 118
Resource indexing, 18, 119
Route computation, 22
Route error, 75
Route reply, 69
Route request, 69
Router, 20, 24
Router buffer, 21
Routing intelligence, 146
Routing table, 20, 21

S
Scalable coding, 150
Scale-free networks, 148
Security, 41
Self-managed networks. *See* Autonomic networks
Shortest-path algorithm, 28, 31
  confirmed paths, 30
  link cost, 30
  optimal distribution tree, 32
  routing table of, 32
  tentative paths, 30
Six degrees of separation property. *See* Small-world networks
Skype, 17
Small-world networks, 147
  bottlenecks, 148
  bridges, 148
  hubs, 148
  nestedness, 149
  randomness, 149
  rich hubs, 148
  robustness of, 148
  stability of, 148
  strong links, 148
  VIP clubs, 149
  weak links, 148
Small-worldness. *See* Small-world networks
Social networks, 5
Spontaneous connectivity, 52
Spontaneous networks, 52
Statistical P2P streaming, 7
Store-match-switch-forward process, 22
Stream chunks, 7
Structured key space, 121
Success rate of requests, 134
Supernode, 118, 128
Synchronous interaction pattern, 15

T
Time-to-live, 106, 134
Topology control, 80
Topology information base, 84
Traffic patterns, 10, 12
Transmission control protocol, 19, 40
Tree-exploration algorithms, 138

U
Ubiquitous applications, 12
Ubiquitous connectivity, 51
Ubiquitous networks, 42
Underlying physical network, 42
Unicast network, 105
Unresponsive networks, 43

V
Video conferencing, 20
Video-on-demand, 19
Virtual interaction paradigm, 15
Virtual link, 95
Virtual networks, 42, 93, 95, 96, 99, 103, 105, 107, 111, 129, 133
  hybrid, 119, 127
  structured, 112, 120, 128, 133, 141, 142
  unstructured, 112, 120, 133, 134, 141
Virtual node, 94
Virtual routing, 95
Voice over IP, 4

W
Web, 1
Windows live synch, 17
World wide web, 1, 40

Z
Zone routing protocol, 90
  bordercast request, 91
  interzone routing protocol, 92
  intrazone routing protocol, 92
  neighbor discovery protocol, 90
  reverse path, 89
  route reply, 88

MIX
Papier aus verantwortungsvollen Quellen
Paper from responsible sources
FSC® C105338

If you have any concerns about our products,
you can contact us on
ProductSafety@springernature.com

In case Publisher is established outside the EU,
the EU authorized representative is:
Springer Nature Customer Service Center GmbH
Europaplatz 3, 69115 Heidelberg, Germany

Printed by Libri Plureos GmbH
in Hamburg, Germany